Developing Blockchain Solutions in the Cloud

Design and develop blockchain-powered Web3 apps on AWS, Azure, and GCP

Stefano Tempesta

Michael John Peña

Developing Blockchain Solutions in the Cloud

Group Product Manager: Preet Ahuja

Publishing Product Manager: Suwarna Rajput

Book Project Manager: Uma Devi

Senior Editor: Romy Dias

Technical Editor: Nithik Cheruvakodan

Copy Editor: Safis Editing

Proofreader: Safis Editing

Indexer: Tejal Daruwale Soni

Production Designer: Alishon Mendonca

Senior DevRel Marketing Executive: Linda Pearlson

DevRel Marketing Coordinator: Rohan Dobhal

First published: April 2024

Production reference: 1280324

Published by Packt Publishing Ltd.

Grosvenor House

11 St Paul's Square

Birmingham

B3 1RB, UK.

ISBN 978-1-83763-017-2

www.packtpub.com

There a lot of people that helped me make this book possible. I want to thank them all and acknowledge their support: Yuliia, Ivan, Clang, Lei, Red, Khel, and Aaron. You are amazing – the best team I could dream of. We, together, left a mark on the Web3 story. This is only the beginning…

– Stefano Tempesta

Contributors

About the authors

Stefano Tempesta is a technologist working at the crossroads of Web2 and Web3 to make the internet a more accessible, meaningful, and inclusive space. Stefano is an ambassador of the use of AI and blockchain technology for good purposes. A former advisor to the Department of Industry and Science, Australia, on the National Blockchain Roadmap, he is cofounder of Aetlas, a decentralized climate action and sustainability network with a mission to source verified carbon units for liquidity and carbon asset monetization. A passionate traveler, a poor musician, and an avid learner of new technologies and (programming) languages, Stefano holds three citizenships and speaks fluent English, Italian, and terrible Ukrainian.

Michael John Peña, an engineer and Microsoft MVP, excels in tech innovation and leadership. As a data partner at Playtime Solutions, he spearheads projects utilizing Azure, big data, and AI, enhancing data-driven decision-making. With roles ranging from CTO to software engineer, MJ's expertise covers web/app development, cloud computing, blockchain, and IoT. His commitment to lifelong learning and sharing knowledge—underscored by his work with start-ups and as a technical advisor—drives industry advancements in finance, construction, and more. MJ values inclusivity and actively fosters diverse, collaborative environments.

About the reviewers

Faisal Ahmed Farooqui is an architect, having more than two decades of experience in the software development industry. He's currently serving as director of engineering and country head at KalSoft. He completed his master's in computer science, postgraduate diploma (computer science), and bachelor's in commerce all at the University of Karachi. To return to society, he taught blockchain in the same Computer Science department at the University of Karachi. In his vast experience in the software industry, he has worked as an architect and developer in multiple technology areas, which include big data platforms, blockchain, DevOps, process automation, AI/ML, microservices, serverless, and many others, particularly provided by AWS, Azure, and others such as GCP, OCI, and Databricks.

Alhamdulillah – all praise to Allah. I'd like to thank my parents, my beautiful wife, my lovely kids – Maryam, a sweet little princess, and Abdul Hadi, a naughty Dirty Bertie, my brothers, sisters, and all family members. Thanks to Akhter Zaib, founding member of KalSoft for being my mentor, teacher, and guide, like an elder brother. Thanks to my previous organizations, bosses, and colleagues for giving me opportunities to learn, grow, and make mistakes.

As a seasoned full-stack developer, **Reigner Garcia Ouano** seamlessly blends creativity and precision in the digital realm. With over six years of experience navigating the intricacies of web development, Reigner Ouano is not only a coding maestro but also a passionate explorer of diverse literary landscapes.

I would like to express my heartfelt gratitude to the author of this book for providing a comprehensive and insightful resource that explores the intricate intersection of blockchain technology and cloud computing. Your dedication to simplifying complex concepts and making them accessible to a broader audience is truly commendable.

Ivan Leskov, a blockchain engineer and security analyst, has an extensive eight-year career in finance, dedicating the last four years specifically to non-traditional, innovative finance realms. Ivan's significant contributions include enhancing various sectors with blockchain technology, notably in certifying collectibles through blockchain in the firearms market and developing DeFi/ReFi protocols. His deep insight into financial systems and innovative blockchain strategies mark him as an influencer, subtly transforming financial landscapes into more efficient, transparent realms. Ivan's work is a testament to his commitment to redefining finance through thoughtful technological progress.

I extend my deepest gratitude to my peers and upcoming experts in the blockchain domain. Your commitment to exploring and advancing this field is invaluable. Your choice to delve into blockchain technology is not just a career path, but a significant step toward shaping a more transparent and secure future. Thank you for your dedication and for joining in this journey to revolutionize our global financial and technological landscapes.

Table of Contents

2

Overview of AWS, Azure, and GCP Services for Blockchain   31

3

DevOps for Cloud-Native Blockchain Solutions   45

Part 2: Deploying and Implementing Blockchain Solutions on AWS

4

Getting Started with Amazon Managed Blockchain   69

5

Hosting a Blockchain Network on Elastic Kubernetes Service   93

6

Building Records with Amazon Quantum Ledger Database   115

Part 3: Deploying and Implementing Blockchain Solutions on Azure

7

9

Leveraging Azure Confidential Ledger   191

Part 4: Deploying and Implementing Blockchain Solutions on GCP

10

Hosting an Ethereum Blockchain Network on Google Cloud Platform   207

Part 5: Exploring Real-World Use Cases and Best Practices

13

14

15

Creating Verifiable Digital Ownership on GCP — 307

16

The Future of Cloud-Native Blockchain — 325

Preface

The adoption of cloud-native blockchain technology is on the rise in various sectors, as companies capitalize on its advantages to stay in sync with technological advancements. This book introduces you to the basic principles of cloud-native blockchain, its benefits, and the obstacles encountered when implementing it in a cloud environment.

Chapter by chapter, you will learn crucial topics, including the development and deployment of robust, scalable Web3 solutions on platforms such as AWS, Azure, and GCP, complemented by practical tutorials and projects for hands-on learning. The book covers the most suitable tools, technologies, practical applications, and recommended practices for Web3, enriching your knowledge about the industry's challenges and opportunities.

The goal of this book is to make you possess a thorough comprehension of how to deploy and implement Web3 applications in the cloud, equipped with the skills and knowledge necessary to create secure and scalable cloud-native blockchain solutions.

Who this book is for

This book is written for cloud developers and DevOps engineers aiming to design and develop cloud-native Web3 applications. It serves as a valuable resource for IT professionals tasked with deploying and managing blockchain solutions in the cloud, as well as business executives assessing the viability of blockchain technology. Additionally, entrepreneurs, students, academics, and enthusiasts keen on discovering the newest advancements in Web3 cloud applications will find it a useful read. A foundational understanding of cloud computing and blockchain principles is advised to fully grasp the expert analysis, practical tutorials, and real-life applications presented.

What this book covers

Chapter 1, *Understanding Cloud-Native and Blockchain*, describes the basics of cloud-native architecture and its relationship with blockchain technology. The chapter will explain what cloud-native means, how it differs from traditional cloud computing, and why it is relevant for blockchain applications. It will also discuss the key benefits of using cloud-native technologies for blockchain, such as scalability, security, and cost-effectiveness.

Chapter 2, *Overview of AWS, Azure, and GCP Services for Blockchain*, provides an overview of the blockchain services offered by AWS, Azure, and GCP. The chapter will provide an understanding of the different services and tools available for building blockchain solutions on each platform. It will also cover the strengths and weaknesses of each platform, as well as the different features and capabilities they offer.

Chapter 3, DevOps for Cloud-Native Blockchain Solutions, explains the principles and practices of DevOps and how they can be applied to cloud-native blockchain solutions. The chapter will provide an understanding of how DevOps can help to build and deploy blockchain solutions more efficiently and effectively, by creating a more collaborative environment between development and operations teams.

Chapter 4, Getting Started with Amazon Managed Blockchain, shows how to get started with Amazon Managed Blockchain on AWS. The chapter will explain the basics of Amazon Managed Blockchain, including how to create and manage a network, invite members, and deploy and manage nodes. The chapter will also cover key considerations for security, scaling, and monitoring when using Amazon Managed Blockchain.

Chapter 5, Hosting a Blockchain Network on Elastic Kubernetes Service, demonstrates how to host a blockchain network on **Elastic Kubernetes Service (EKS)** on AWS. The chapter will explain the basics of EKS, including how to create and manage a cluster and how to deploy a blockchain network on EKS. The chapter will also cover key considerations for security, scaling, and monitoring when using EKS for blockchain.

Chapter 6, Building Records with Amazon Quantum Ledger Database, explores the use of Amazon **Quantum Ledger Database (QLDB)** on AWS for blockchain applications. The chapter will explain the basics of QLDB, including its features and benefits for blockchain applications, and how to create and manage a QLDB instance. The chapter will also cover key considerations for data modeling and querying on QLDB.

Chapter 7, Hosting a Corda DLT Network on Azure Kubernetes Service, describes how to host a blockchain network on **Azure Kubernetes Service (AKS)** on Azure. The chapter will explain the basics of AKS, including how to create and manage a cluster and how to deploy a blockchain network on AKS. The chapter will also cover key considerations for security, scaling, and monitoring when using AKS for blockchain.

Chapter 8, Using the Ledger Features of Azure SQL, dives into use cases for the ledger features of Azure SQL to implement blockchain solutions. The chapter will provide an overview of the ledger features of Azure SQL, including its key features and benefits, and will explain how it can be used to implement blockchain solutions.

Chapter 9, Leveraging Azure Confidential Ledger, is all about Microsoft Azure Confidential Ledger, a confidential and secure ledger that provides a transparent and verifiable history of all changes to the data stored in a ledger. The chapter will provide an overview of Microsoft Azure Confidential Ledger, including its features and benefits, and will explain how it can be used to implement blockchain solutions.

Chapter 10, Hosting an Ethereum Blockchain Network on Google Cloud Platform, shows how to host a blockchain network on GCP using **Google Kubernetes Engine (GKE)**. The chapter will provide a step-by-step guide for setting up a blockchain network on GKE, including instructions for deploying the network, configuring the nodes, and managing the network.

Chapter 11, *Getting Started with Blockchain Node Engine*, focuses on Blockchain Node Engine, a service provided by GCP for building and deploying blockchain applications. The chapter will provide an overview of Blockchain Node Engine, including its features and benefits, and will explain how it can be used to implement blockchain solutions.

Chapter 12, *Analyzing On-Chain Data with BigQuery*, is dedicated to Google BigQuery, a data analysis service provided by GCP, to analyze on-chain data in their blockchain applications. The chapter will provide an overview of BigQuery, including its features and benefits, and will explain how it can be used to analyze on-chain data in a blockchain network.

Chapter 13, *Building a Decentralized Marketplace on AWS*, is our first hands-on lab on how to build a decentralized marketplace using AWS services. The chapter will provide practical instructions that will guide you through the process of building a decentralized marketplace on AWS, including instructions for deploying the network, creating the marketplace application, and integrating the marketplace with other AWS services.

Chapter 14, *Developing a Decentralized Voting Application on Azure*, is a second lab for developing a decentralized voting application using Azure services. The chapter will provide a hands-on guide through the process of developing a decentralized voting application on Azure, including instructions for deploying a network, creating a voting application, and integrating it with other Azure services.

Chapter 15, *Creating Verifiable Digital Ownership on GCP*, is our third lab on how to create verifiable digital ownership records using GCP services. The chapter will provide a hands-on lab that guides you through the process of creating verifiable digital ownership records on GCP, including instructions for deploying a network, creating ownership records, and integrating them with other GCP services.

Chapter 16, *The Future of Cloud-Native Blockchain*, closes this book with a summary of the key points covered throughout the previous chapters, and then discusses the future of cloud-native blockchain. This chapter will provide insights into the current state and future trends of cloud-native blockchain, including the adoption of blockchain for good humanitarian purposes, the challenges, and its opportunities.

To get the most out of this book

Software/hardware covered in the book	Operating system requirements
A subscription to each of the cloud services utilized in the book – AWS, Azure, and GCP.	Any
A development tool of preference for smart contracts and other scripts. VS Code is recommended.	Any

If you are using the digital version of this book, we advise you to type the code yourself or access the code from the book's GitHub repository (a link is available in the next section). Doing so will help you avoid any potential errors related to the copying and pasting of code.

Download the example code files

You can download the example code files for this book from GitHub at `https://github.com/PacktPublishing/Developing-Blockchain-Solutions-in-the-Cloud`. If there's an update to the code, it will be updated in the GitHub repository.

We also have other code bundles from our rich catalog of books and videos available at `https://github.com/PacktPublishing/`. Check them out!

Conventions used

There are a number of text conventions used throughout this book.

`Code in text`: Indicates code words in text, database table names, folder names, filenames, file extensions, pathnames, dummy URLs, user input, and Twitter handles. Here is an example: "The `PeerNode1AZ` and `PeerNode2AZ` parameters identify the selected AWS region for deployment of the two peer nodes."

A block of code is set as follows:

```
response = cloudwatch_client.put_dashboard(
    DashboardName=dashboard_name,
    DashboardBody=json.dumps(dashboard_body)
)
```

When we wish to draw your attention to a particular part of a code block, the relevant lines or items are set in bold:

```
cloudwatch_client = boto3.client('cloudwatch')
alarm_name = 'AMB_CPU_Utilization_Alarm'
namespace = 'AWS/ManagedBlockchain'
metric_name = 'CPUUtilization'
threshold = 70.0  # Set your threshold
```

Any command-line input or output is written as follows:

```
$ pip install boto3
```

Bold: Indicates a new term, an important word, or words that you see on screen. For instance, words in menus or dialog boxes appear in **bold**. Here is an example: "Click on the **Private networks** tab and **Create private network** button in the AMB console."

> **Tips or important notes**
> Appear like this.

Get in touch

Feedback from our readers is always welcome.

General feedback: If you have questions about any aspect of this book, email us at `customercare@packtpub.com` and mention the book title in the subject of your message.

Errata: Although we have taken every care to ensure the accuracy of our content, mistakes do happen. If you have found a mistake in this book, we would be grateful if you would report this to us. Please visit `www.packtpub.com/support/errata` and fill in the form.

Piracy: If you come across any illegal copies of our works in any form on the internet, we would be grateful if you would provide us with the location address or website name. Please contact us at `copyright@packt.com` with a link to the material.

If you are interested in becoming an author: If there is a topic that you have expertise in and you are interested in either writing or contributing to a book, please visit `authors.packtpub.com`.

Share your thoughts

Once you've read *Developing Blockchain Solutions in the Cloud*, we'd love to hear your thoughts! Scan the QR code below to go straight to the Amazon review page for this book and share your feedback.

`https://packt.link/r/1837630178`

Your review is important to us and the tech community and will help us make sure we're delivering excellent quality content.

Download a free PDF copy of this book

Thanks for purchasing this book!

Do you like to read on the go but are unable to carry your print books everywhere?

Is your eBook purchase not compatible with the device of your choice?

Don't worry, now with every Packt book you get a DRM-free PDF version of that book at no cost.

Read anywhere, any place, on any device. Search, copy, and paste code from your favorite technical books directly into your application.

The perks don't stop there, you can get exclusive access to discounts, newsletters, and great free content in your inbox daily

Follow these simple steps to get the benefits:

1. Scan the QR code or visit the link below

```
https://packt.link/free-ebook/9781837630172
```

2. Submit your proof of purchase
3. That's it! We'll send your free PDF and other benefits to your email directly

Part 1: Introduction to Cloud-Native Blockchain

This part provides an understanding of cloud-native technology and blockchain, and an overview of the AWS, Azure, and GCP services available for implementing blockchain solutions. It will also provide insights on DevOps practices for cloud-based blockchain solutions.

This part includes the following chapters:

- *Chapter 1, Understanding Cloud-Native and Blockchain*
- *Chapter 2, Overview of AWS, Azure, and GCP Services for Blockchain*
- *Chapter 3, DevOps for Cloud-Native Blockchain Solutions*

1

Understanding Cloud-Native and Blockchain

In today's rapidly evolving technological landscape, it is vital to stay ahead by understanding and adopting key technologies that are driving innovation and reshaping industries. This chapter will provide you with foundational knowledge of cloud-native and blockchain technologies, their benefits, and how they can be effectively combined to deliver solutions across multiple industries. As you progress through this chapter, you will be exposed to blockchain services in the major cloud service providers—AWS, Azure, and GCP—and make informed decisions about implementing cloud-native blockchain solutions. We will explore each technology in-depth, and provide real-world examples to demonstrate their applications and potential impact.

We will cover the following main topics in this chapter:

- Blockchain basics – Distributed ledgers and cryptography
- Introduction to cloud-native technology
- Benefits and limitations of cloud-native blockchain
- Key considerations for cloud-native blockchain implementation

Blockchain basics – Distributed ledgers and cryptography

Blockchain is a decentralized, distributed ledger technology that enables the secure recording, storage, and verification of transactions across a network of computers. It consists of a chain of blocks, where each block contains a list of transactions. These blocks are linked together using cryptographic techniques, forming a chronological chain.

One of the key problems that blockchain addresses is the issue of trust in traditional centralized systems. In traditional systems, such as banking or centralized databases, there is a reliance on intermediaries, such as banks or clearinghouses, to facilitate and validate transactions. This reliance can lead to issues such as fraud, data manipulation, censorship, and single points of failure.

Blockchain technology mitigates these issues by decentralizing the control and storage of data. Transactions on a blockchain are verified and recorded by a network of nodes, and once added to the blockchain, they are immutable and tamper-resistant. This decentralized and transparent nature of blockchain technology removes the need for intermediaries and creates a more trustless environment for transactions to occur.

Blockchain also addresses the problem of data integrity and security. The cryptographic techniques that are used in blockchain ensure that data stored on the ledger cannot be altered or tampered with without consensus from the network participants. This makes blockchain particularly suitable for applications where data integrity and security are paramount, such as financial transactions, supply chain management, voting systems, and identity management.

Myriads of books and online resources have already been written on introducing blockchain and the relevant technologies and frameworks that run along a blockchain network. Before progressing with this book, we recommend the following resources for deepening your understanding of cryptocurrency and tokens, the type and use of wallets, addresses, and cryptographic concepts, such as private and public keys, and hash values:

- *Mastering Blockchain*: `https://www.packtpub.com/product/mastering-blockchain-third-edition/9781839213199`

- *Complete Cryptocurrency and Blockchain Course*: `https://www.packtpub.com/product/complete-cryptocurrency-and-blockchain-course-learn-solidity-video/9781839211096`

- *Cryptography Algorithms*: `https://www.packtpub.com/product/cryptography-algorithms/9781789617139`

- *Hands-On Cryptography with Python*: `https://www.packtpub.com/product/hands-on-cryptography-with-python/9781789534443`

Types of blockchain

Blockchain operates as a form of **Distributed Ledger Technology** (**DLT**) that enables various stakeholders to safely collaborate and oversee an ever-expanding series of data entries, termed blocks. Every block encapsulates a group of activities, a time marker, and a nod to the preceding block. Together, these blocks create a resilient archive of all engagements undertaken on the platform.

One of the standout attributes of blockchain is its non-centralized design. This entails that the data isn't dominated by a singular authority; instead, every member on the platform possesses a duplicate of the full ledger. Such a distributed model negates the dependency on centralized regulators or middlemen, curtailing possibilities of manipulation, deceit, and singular breakdowns.

Encryption is pivotal in preserving the trustworthiness of the blockchain. Each block carries a distinct cryptographic signature, conceived from the block's details and the signature of its predecessor. This setup renders it exceedingly challenging to modify a block's data without influencing the blocks that follow, guaranteeing the permanence of the blockchain.

Blockchains can be classified into three primary categories: public, private, and consortium. Each serves its purpose with distinct features and preferred applications.

Public blockchains

Public blockchains operate as open platforms accessible to any individual willing to join. Bitcoin and Ethereum are classic representations of this type. These blockchains pride themselves on deep-rooted decentralization and formidable security measures. However, their intricate validation procedures, such as **Proof of Work (PoW)**, can sometimes restrict their speed and scalability. Public blockchains excel in environments that demand trust, openness, and resistance to external controls, such as with **decentralized apps (dapps)** and virtual currencies.

Private blockchains

Private blockchains, also known as permissioned blockchains, in contrast, are limited-access platforms that are often confined to specific entities or organizations. They are tailored for rapid transactions and superior scalability, presenting an edge over public blockchains in these aspects. However, this comes at the cost of shedding some of the complete decentralization inherent in public counterparts. For tasks prioritizing confidentiality, oversight, and swift operations, such as within organizational collaborations, supply chain tracking, or internal documentation, private blockchains emerge as the go-to choice.

Here are some examples of private blockchain platforms:

- **Hyperledger Fabric**: Developed by the Linux Foundation, Hyperledger Fabric is designed for enterprise use. It enables organizations to build scalable blockchain applications with a high degree of privacy, performance, and customization. Industries such as finance, healthcare, and supply chain use Hyperledger Fabric for applications such as asset tracking, secure transactions, and compliance.

- **R3 Corda**: Corda is a distributed ledger platform designed specifically for the financial industry but has applications in other sectors as well. It allows for the development of interoperable blockchain networks where transaction privacy is maintained, enabling businesses to transact directly and in strict privacy.

- **Quorum**: Initially developed by J.P. Morgan, Quorum is an Ethereum-based private blockchain. It's designed for processing private transactions within a permissioned group of known participants. Quorum is used in sectors such as finance for settlement, payments, and other financial services where transaction privacy is crucial.

- **Ripple (XRP Ledger for Private Use)**: While Ripple's XRP Ledger is public, Ripple has developed a private version of it for central banks to issue and manage digital currencies. This private blockchain solution offers the transaction privacy and control that financial institutions require.

- **Multichain**: Multichain is a platform that allows users to establish private blockchains for financial transactions, asset management, and other applications. It provides the tools for creating and deploying blockchain applications with a focus on privacy, control, and scalability.

Consortium blockchains

Consortium blockchains, often referred to as federated blockchains, interweave elements from both public and private types. They function under the stewardship of a group of trusted entities, rather than a single overarching body. Striking a balance, consortium blockchains marry the transparency of public chains with the control advantages of private ones. They are especially apt for situations that entail collaboration across different organizations but also necessitate a degree of privacy and control, such as in cross-border banking transactions or shared healthcare record systems.

Here are some examples of consortium blockchain platforms and their applications:

- **Energy Web Foundation (EWF)**, `https://www.energyweb.org/`: EWF has developed the Energy Web Chain, a public, enterprise-grade blockchain platform designed for the energy sector's specific needs. It supports a consortium of energy companies working together to develop dapps that can drive cleaner, more efficient, and inclusive energy systems worldwide.

- **Blockchain Insurance Industry Initiative (B3i)**: B3i was a consortium of insurance companies that came together to explore and implement blockchain solutions. B3i used blockchain technology to make insurance transactions more efficient, transparent, and customer-friendly. B3i terminated operations in 2022.

- **we.trade**: A consortium blockchain developed by a group of 12 major banks in Europe, we.trade was designed to simplify and secure international trade transactions for companies. The platform leveraged smart contracts to ensure that all parties in the supply chain could meet their obligations, offering a streamlined, secure process for trade finance. we.trade closed in 2022, but its case study by IBM is still accessible at `https://www.ibm.com/case-studies/wetrade-blockchain-fintech-trade-finance`.

These few examples illustrate how consortium blockchains are being used across various industries to improve processes, enhance security, and facilitate collaboration between different organizations.

Consensus mechanisms – Ensuring trust and security

A core attribute of blockchain technology is its proficiency in ensuring agreement amid a distributed set of nodes. These consensus models are computational strategies that validate uniform agreement among network nodes regarding the blockchain's status and the legitimacy of upcoming transactions. Various consensus strategies exist, each tailored with its unique benefits and limitations. Notable consensus models encompass the following:

- **PoW**: Adopted primarily by Bitcoin and similar digital assets, PoW mandates participants (or *miners*) to employ robust computing resources to calculate intricate algorithms. When a miner cracks a challenge, in the case of Bitcoin, for example, finding a hash value that matches the requirements of the network, they suggest integrating a new block into the blockchain. The network's other nodes then scrutinize the block's dealings, culminating in an agreement to either approve or decline the block. While PoW assures formidable security, it is power-hungry and might lean toward concentration:

Figure 1.1 – PoW in Bitcoin

The preceding AI-generated image visually explains the PoW consensus mechanism in Bitcoin. It depicts how miners use high-powered computers to compete in solving complex mathematical puzzles, with the successful miner unlocking a new block through the discovery of the correct hash. This process is symbolized by a golden key unlocking a block. The preceding figure also shows a network of nodes connected by digital links, highlighting the decentralized validation and addition of the newly mined block to the blockchain. The energy consumption involved in mining is represented by electricity symbols around the miners' computers, providing a comprehensive overview of Bitcoin mining and the PoW system.

- **Proof of Stake (PoS)**: PoS emerges as a response to PoW's limitations. Instead of expending computational might, PoS hinges on the quantity of digital currency a user possesses (their *holdings*) to gauge their likelihood of initiating a new block. Those with heftier holdings usually have elevated chances of being elected for block validations. PoS is considerably less power-consuming than PoW, though it might still edge toward central tendencies:

Figure 1.2 – PoS in Ethereum

The preceding AI-generated image has been created to visually explain the PoS consensus mechanism in Ethereum. It illustrates validators staking Ethereum tokens, the selection process symbolized by a digital scale, and the decentralized network of nodes. The energy efficiency of PoS is also conveyed through green energy symbols, providing a clear, engaging, and informative view of Ethereum's PoS system.

- **Delegated Proof of Stake (DPoS)**: As an offshoot of PoS, DPoS injects an element of representative choice into the agreement process. Here, the most substantial stakeholders empower specific, trustworthy nodes with their voting privileges. These chosen nodes then authenticate dealings and craft new blocks. DPoS can expedite transactions and enhance scalability, yet might be susceptible to collective manipulations and centralization.

- **Practical Byzantine Fault Tolerance (PBFT)**: PBFT is a consensus model tailored for permissioned blockchain networks where every node is identifiable and trustworthy. Within this structure, one node gets designated as the primary proposer for a fresh block. The remaining nodes in the system affirm the block through message exchanges. After a dominant majority concurs on the block's authenticity, it joins the blockchain. While PBFT assures swift transactions and robust security, it might not be the best fit for expansive, open blockchain systems.

Smart contracts – Programmable logic on the blockchain

Smart contracts represent digital contracts where the conditions of an agreement are embedded directly into the programming. Operating on blockchain platforms, they facilitate automatic and trust-free completion of deals between entities, eliminating the reliance on middlemen. These contracts can be tailored to execute a range of functions, such as handling digital asset transfers, streamlining business operations, or activating contract stipulations upon meeting certain criteria.

Ethereum pioneered the integration of smart contracts into blockchain technology, though now, multiple other blockchain systems, such as Cardano, Polkadot, and Tezos, also incorporate smart contract capabilities.

The adaptability of smart contracts opens avenues for their usage in diverse sectors, encompassing finance, logistics, insurance, property dealings, and dapps.

From a programming perspective, it's important to emphasize that conventional programming and smart contract development are two distinct paradigms within the software development landscape, each with its unique characteristics, use cases, and challenges. Let's have a look at some key differences between conventional programming and smart contract development:

Programming paradigm	Conventional	Smart contract
Execution Environment	Applications are usually executed in a diverse range of environments, from personal computers and servers to cloud platforms. These environments can vary greatly in terms of their operating systems, hardware configurations, and network connections.	Smart contracts are executed on a blockchain platform, such as Ethereum. This environment is deterministic, meaning that the execution of the contract will always produce the same output given the same initial state and inputs, across all nodes in the network.
Language and Tools	There is a wide range of programming languages, frameworks, and tools available, each suited to different tasks, from web and mobile app development to data analysis and system programming.	Development is typically done in domain-specific languages such as Solidity (for Ethereum) or Java (for Hyperledger Fabric). These languages are designed to facilitate the creation of blockchain-based applications but might have limitations or unique features compared to more general-purpose languages.

Update and Maintenance	Conventional software can be updated or patched as needed. Developers can push updates to fix bugs, add features, or improve performance, and users can usually apply these updates at their convenience.	Once deployed, a smart contract is immutable; it cannot be changed or updated. If a bug is found or an update is required, a new contract must be deployed, and the state and assets controlled by the old contract may need to be manually migrated to the new one.
State Management and Transactions	State management is handled within the application's environment, and transactions (if applicable) are managed by external systems such as databases or payment processors.	State management and transactions are intrinsic to the blockchain platform. Smart contracts not only manage the state of applications but also execute transactions that are transparent, traceable, and irreversible, within the blockchain network.
Consensus Mechanism	It does not inherently involve a consensus mechanism for decision-making. The application's behavior is determined by its code and the inputs it receives.	Execution and the validity of transactions are subject to a consensus mechanism (for example, PoW or PoS) among the participants in the blockchain network. This ensures agreement on the state of the distributed ledger and the results of smart contract executions.
Security Implications	Security is crucial in all programming, but the impact of vulnerabilities can vary. Some issues can be patched before they are exploited, or their impact can be mitigated through various means.	Security is paramount and potentially more challenging due to the immutable nature of smart contracts. Vulnerabilities in smart contracts can lead to irreversible loss or theft of digital assets. This necessitates rigorous testing, audits, and formal verification processes before deployment.

Table 1.1 – Comparison between conventional and contract-oriented development

In summary, while both conventional programming and smart contract development share the fundamental principles of software development, they diverge significantly in their execution environments, security considerations, update mechanisms, and the ways they handle transactions and state management. These differences necessitate distinct approaches to development, testing, and deployment in each domain.

dapps – Building on the blockchain

dapps are applications that are built on top of blockchain platforms, utilizing smart contracts and decentralized storage to create trustless, transparent, and censorship-resistant services. dapps leverage the unique features of blockchain technology, such as immutability, decentralization, and tokenization, to deliver innovative solutions and disrupt traditional industries.

Examples of dapps include **Decentralized Finance (DeFi)** platforms, decentralized marketplaces, **Decentralized Autonomous Organizations (DAOs)**, and **Non-Fungible Tokens (NFTs)**.

Tokenization – Creating digital assets on the blockchain

Tokenization is the process of representing real-world assets or rights on the blockchain in the form of digital tokens. These tokens can represent anything from digital currencies and financial instruments to physical assets, such as real estate, art, or commodities.

There are two main types of tokens in the blockchain ecosystem:

- **Fungible tokens**: Fungible tokens are interchangeable and have a consistent value across all instances. They are commonly used to represent digital currencies, such as Bitcoin and Ether, as well as other digital assets, such as utility tokens or security tokens.

- **NFTs**: NFTs are unique, indivisible, and non-interchangeable tokens that represent ownership of a specific digital or physical asset. Each NFT has a unique identifier, which distinguishes it from other tokens. NFTs have gained significant popularity in the world of digital art, collectibles, and virtual goods as they provide a way to prove the authenticity and ownership of these assets on the blockchain.

> **Cryptocurrencies and tokens**
>
> Cryptocurrencies and tokens, such as NFTs, are both digital assets, but they serve different purposes and operate on different principles. For a start, cryptocurrency works at the network level, while tokens are governed by smart contracts. Cryptocurrencies such as Bitcoin or Ether are fungible, meaning that each unit is interchangeable with another unit of the same value. For example, one Bitcoin is equal in value to any other Bitcoin. NFTs, on the converse, are non-fungible, meaning that each token is unique and cannot be replaced with another token of the same value. In general, while both cryptocurrencies and tokens are digital assets, they serve different purposes, have different properties, and operate on different blockchain standards. Cryptocurrencies are primarily used for financial transactions, while tokens represent ownership or authenticity of digital or physical items.

Tokenization has the potential to revolutionize various industries by enabling the creation, trading, and management of digital assets in a secure, transparent, and decentralized manner.

Scalability, interoperability, and privacy – Key challenges and innovations in blockchain

As blockchain technology continues to evolve, several key challenges must be addressed to unlock its full potential. These challenges include scalability, interoperability, and privacy.

Scalability

Scalability is the ability of a blockchain network to handle an increasing number of transactions without compromising performance. Many public blockchains, such as Bitcoin and Ethereum, face scalability issues due to their resource-intensive consensus mechanisms and limited transaction throughput.

Various solutions are being developed to improve the scalability of blockchain networks, including the following:

- **Layer 2 solutions**: These are off-chain protocols that process transactions outside the main blockchain, reducing the load on the network. Examples of layer 2 solutions include the Lightning Network for Bitcoin and Polygon (and many others) for Ethereum.

  - The **Bitcoin Lightning Network** (`https://lightning.network/`) is a layer-two scaling solution that's designed to address the scalability and transaction throughput limitations of the Bitcoin blockchain. It is built on top of the Bitcoin protocol and operates as a decentralized network of payment channels.

  - **Polygon** (`https://polygon.technology/`) is a Layer 2 scaling solution for Ethereum that aims to address the network's scalability issues by providing faster and cheaper transactions. It achieves this by using sidechains, Plasma chains, and other scaling techniques to offload transactions from the Ethereum mainnet.

- **Sharding**: This is a technique that involves dividing the blockchain into smaller, parallel chains (shards) that can process transactions independently, thus increasing the overall transaction throughput.

- **New consensus algorithms**: Alternative consensus mechanisms, such as PoS and DPoS, can offer improved scalability compared to PoW.

Interoperability

Interoperability refers to different blockchain platforms' capability to engage and establish communication with each other. At present, the majority of blockchains function independently, constraining opportunities for collaboration and information sharing across chains.

Multiple initiatives focus on bolstering the mutual communication among blockchain platforms. Here are some examples:

- **Cross-chain bridges**: These are structured systems that support the movement of assets and information from one blockchain network to another. Notable instances are the **Wrapped Bitcoin (WBTC)** token, enabling Bitcoin's operation on the Ethereum network, and the **Interledger Protocol (ILP)**, which simplifies payments across chains.

- **Blockchain-agnostic platforms**: Platforms of this kind empower developers to craft tools that can engage with diverse blockchain networks. Some of these frameworks include Polkadot, Cosmos, and Avalanche.

Privacy

Preserving confidentiality remains pivotal for numerous blockchain applications since actions on open blockchains can often be viewed and traced. This transparency becomes problematic for scenarios necessitating discreet information storage and operations on the blockchain.

Several innovations aimed at bolstering privacy are in development:

- **Zero-knowledge proofs**: These cryptographic methods allow an entity to confirm the authenticity of a claim without disclosing specifics about the claim. Techniques such as zk-SNARKs and zk-STARKs are prominent instances of such proofs.

- **Confidential transactions**: These are protocols that are designed to obscure transaction values on the blockchain, ensuring heightened privacy while still retaining traceability. The Mimblewimble protocol, adopted by cryptocurrencies such as Grin and Beam, is a prominent example of shielded transactions.

- **Private and consortium blockchains**: As previously discussed, selective-access blockchains, both private and consortium types, provide an enhanced level of confidentiality when juxtaposed with public blockchains as they limit entry to a particular set of trusted stakeholders.

By understanding these core concepts, we now have a solid foundation in blockchain technology and its potential to revolutionize various industries. Now, we can delve deeper into the benefits and limitations of building cloud-native decentralized applications that leverage blockchain technology.

Introduction to cloud-native technology

In this section, we will explore the concept of cloud-native technology, its underlying principles, and the benefits it brings to the modern software development landscape. We will discuss key cloud-native concepts, such as containerization, orchestration, and microservices, and provide examples of how these technologies enable the development of flexible, scalable, and resilient applications.

Defining cloud-native technology

Cloud-native technology refers to a software development approach that leverages the advantages of cloud computing to build, deploy, and manage applications. This approach prioritizes flexibility, scalability, and resilience by leveraging modern technologies and techniques such as containerization, microservices, **Continuous Integration and Continuous Delivery (CI/CD)**, and **Infrastructure as Code (IaC)**.

Cloud-native applications are designed to run on distributed and scalable infrastructure, allowing them to adapt to changing workloads and requirements quickly. This enables organizations to respond to market changes more rapidly, improve resource utilization, and optimize the **Total Cost of Ownership (TCO)** of their applications.

Key principles of cloud-native technology

There are several key principles and technologies associated with cloud-native applications. These include containerization, orchestration, microservices, CI/CD, and IaC.

Containerization

Containerization is a lightweight virtualization technology that allows applications and their dependencies to be packaged into isolated, portable containers. Containers can run on any platform that supports the container runtime, providing a consistent environment for development, testing, and production.

Containers provide several advantages over traditional virtualization techniques, such as reduced overhead, faster startup times, and improved resource utilization. Examples of popular containerization technologies include **Docker** and **containerd**.

Orchestration

Orchestration is the process of managing and automating the deployment, scaling, and operation of containerized applications. Container orchestration platforms, such as **Kubernetes**, **Apache Mesos**, and **Docker Swarm**, provide a framework for defining, deploying, and managing containerized applications at scale.

Orchestration platforms offer various features to support cloud-native applications, including service discovery, load balancing, rolling updates, auto-scaling, and self-healing. These features enable organizations to build and maintain applications that are highly available, scalable, and resilient.

Microservices

Microservices is a software architecture pattern that involves breaking down an application into small, loosely coupled components or services. Each service is responsible for a specific piece of functionality and can be independently developed, deployed, and scaled. This approach improves the modularity, maintainability, and scalability of applications, as each service can be updated or replaced without impacting the entire system.

Microservices can be implemented using a variety of languages, frameworks, and platforms, allowing organizations to choose the best tools for their specific needs. This flexibility also enables teams to adopt new technologies and methodologies more easily as each service can evolve independently of the others.

CI/CD

CI/CD is a set of practices that involve the automatic building, testing, and deployment of application code changes. This approach ensures that new features, bug fixes, and other changes are integrated and delivered to users quickly and reliably.

Continuous integration involves the automatic compilation and testing of code changes, ensuring that the application remains in a releasable state at all times. Continuous delivery extends this process to include the automatic deployment of changes to production, minimizing the risk of human error and ensuring a consistent, repeatable deployment process.

CI/CD practices are a core component of cloud-native development as they enable organizations to respond to market changes more rapidly and maintain high-quality applications.

IaC

IaC is a practice that involves managing and provisioning infrastructure components, such as networks, storage, and compute resources, using code and automation. IaC allows organizations to define their infrastructure requirements using code templates, which can then be automatically deployed and managed using tools such as **Terraform**, **Azure Resource Manager**, **AWS CloudFormation**, and **Google Cloud Deployment Manager**.

By treating IaC, organizations can improve the repeatability, reliability, and consistency of their infrastructure deployments. Additionally, IaC enables organizations to adopt version control and collaborative development practices for their infrastructure, ensuring that changes are tracked and reviewed before being applied.

Comparing traditional cloud computing and cloud-native technology

To better understand the advantages of cloud-native technology, it's helpful to compare it with traditional cloud computing approaches. While both methods involve hosting applications and infrastructure on remote servers, they differ significantly in terms of architecture, development practices, and overall philosophy.

Architecture

Traditional cloud computing often involves monolithic applications running on **Virtual Machines** (**VMs**) or physical servers. In this model, applications are typically built as a single, large unit that contains all the required functionality. This can make the application difficult to maintain and scale as changes to one part of the system can impact the entire application.

In contrast, cloud-native applications are built using a microservices architecture, where the application is broken down into small, loosely coupled components. This approach allows individual components to be developed, deployed, and scaled independently, making the overall system more modular and easier to manage.

Development practices

Traditional cloud computing often relies on manual processes and ad hoc scripting for deploying and managing applications. This can lead to inconsistencies and errors, as well as slow release cycles and increased risk of downtime.

Cloud-native technology emphasizes automation and the use of modern development practices such as CI/CD, IaC, and containerization. These practices enable rapid, reliable, and repeatable deployments, as well as improved collaboration and visibility across development teams.

Infrastructure management

In traditional cloud computing, infrastructure management is often treated as a separate concern from application development. This can result in siloed teams and processes, as well as increased complexity and risk.

With cloud-native technology, infrastructure is treated as an integral part of the application life cycle. Using IaC and orchestration tools, developers can define, deploy, and manage infrastructure components alongside application code, ensuring a consistent and repeatable process.

In the next section, we will discuss the fundamentals of blockchain and related technologies.

Benefits and limitations of cloud-native blockchain

In this section, we'll examine the advantages and drawbacks of adopting cloud-native blockchain solutions. We will discuss how cloud-native technologies can enhance the performance, security, and cost-efficiency of blockchain implementations and explore the potential trade-offs and limitations.

Scalability – Adapting to changing workloads

One of the key benefits of cloud-native blockchain solutions is their ability to scale in response to changing workloads. Scalability is essential for blockchain applications that need to handle a growing number of transactions or users. By leveraging cloud-native technologies such as containerization, microservices, and auto-scaling, blockchain implementations can dynamically adapt to meet increasing demands. Consider, though, that cloud technologies are designed for elasticity and scalability, whereas blockchain networks don't necessarily address this technical requirement. As we will see later in this chapter, scalability and performance for blockchain networks can be met with different strategies at the protocol level, including transaction rollups, state channels, and off-chain storage.

Auto-scaling

Auto-scaling is a feature provided by many cloud providers, such as AWS, Azure, and GCP, that automatically adjusts the number of resources allocated to an application based on its current workload. This capability enables cloud-native blockchain implementations to dynamically scale up or down to meet fluctuating demands, ensuring optimal performance and resource utilization.

Security – Protecting data and infrastructure

Security is a critical concern for blockchain applications as they often involve sensitive data and transactions. Cloud-native technologies offer various security features that can enhance the security posture of a blockchain implementation.

Isolation and sandboxing

Containerization provides a natural layer of isolation between applications and the underlying infrastructure. By running each application component in a separate container, the potential impact of a security breach is limited to the compromised container, reducing the risk of lateral movement within the system.

Sandboxing is a technique that restricts the access and permissions of an application component, limiting its potential to cause harm. Cloud-native blockchain solutions can leverage sandboxing to confine smart contracts and other components, reducing the risk of vulnerabilities or malicious code compromising the entire system.

Built-in security features and compliance

Cloud providers offer a wide range of built-in security features that can be leveraged by cloud-native blockchain implementations. These features include data encryption, identity and access management, network security, and security monitoring and logging.

Additionally, many cloud providers are compliant with various industry standards and regulations, such as GDPR, HIPAA, and PCI-DSS. By utilizing cloud-native blockchain solutions, organizations can benefit from these compliance measures without having to implement them independently.

Cost-effectiveness – Optimizing resource utilization

Cloud-native blockchain solutions can provide significant cost savings compared to traditional, on-premises implementations. By leveraging the pay-as-you-go pricing models offered by cloud providers, organizations can optimize resource utilization and reduce costs associated with hardware, maintenance, and energy consumption:

- **Pay-as-you-go pricing**: Pay-as-you-go pricing models allow organizations to pay only for the resources they consume, instead of committing to fixed hardware and infrastructure costs. This model enables cost optimization as resources can be dynamically scaled to match changing workloads, avoiding over-provisioning and under-utilization.

- **Reduced infrastructure and maintenance costs**: By adopting cloud-native blockchain solutions, organizations can eliminate the need to purchase, maintain, and upgrade physical hardware and infrastructure. This approach not only reduces upfront capital expenditures but also minimizes ongoing maintenance and energy costs.

- **Cost optimization tools and techniques**: Cloud providers offer various cost optimization tools and techniques that can help organizations fine-tune their resource usage and spending. These tools include budget and cost management features, as well as recommendations for optimizing resource allocation and performance. By leveraging these tools, organizations can ensure that their cloud-native blockchain implementations remain cost-effective and efficient.

Limitations and trade-offs of cloud-native blockchain solutions

While cloud-native blockchain solutions offer numerous benefits, there are also potential trade-offs and limitations to consider. Some of these challenges include data sovereignty, vendor lock-in, and network latency.

Data governance

The idea behind data governance is that data must adhere to the legal frameworks of the nation where it's housed. When companies integrate blockchain systems into a public cloud setting, they might encounter the challenge of meeting data governance standards. These standards can be intricate, with nuances from one region to the next.

In response to these challenges, companies can opt to launch their blockchain infrastructures on cloud platforms that operate data centers in the preferred legal territories. Additionally, they might leverage mixed or diversified cloud models, ensuring data dispersion across various geographical zones.

Vendor lock-in

Vendor lock-in is a potential concern when adopting cloud-native technologies as organizations may become reliant on a specific cloud provider's infrastructure, services, and tools. This dependency can make it challenging to switch providers or migrate back to on-premises solutions.

To mitigate vendor lock-in risks, organizations can adopt multi-cloud strategies, utilize open source technologies, and implement standardized APIs and data formats that facilitate interoperability between different platforms.

Network latency

Network latency can be a concern for some blockchain applications, especially those that require real-time processing or low-latency interactions. While cloud providers offer various techniques and services to optimize network performance, such as **Content Delivery Networks (CDNs)** and dedicated network connections, latency may still be higher compared to on-premises implementations.

Organizations should carefully assess their specific latency requirements and choose the appropriate cloud-native architecture and services to meet their needs.

Key considerations for cloud-native blockchain implementation

In this section, we will explore the key considerations and best practices for implementing cloud-native blockchain solutions on AWS, Azure, and GCP. By understanding these factors, organizations can make informed decisions and ensure the successful deployment and management of their blockchain applications in a cloud-native environment.

Choosing the right blockchain framework

Determining the right blockchain platform is paramount for executing a cloud-native blockchain initiative. Given that each platform possesses its distinct attributes, strengths, and challenges, it's crucial to pinpoint the one that resonates most with an organization's distinct needs. There are many blockchain networks in the market, and it would be impossible to cover all of them in depth. In the next few sections, we'll look at the three platforms that better represent a broad utilization of permissionless and permissioned blockchain technologies for building decentralized apps: Ethereum, Hyperledger Fabric, and Corda.

Ethereum

Ethereum presents itself as a decentralized, open source platform endorsing smart contracts and facilitating the birth of dapps. Let's look at some of its standout features:

- An open, unrestricted blockchain network

- Endorsement for smart contracts primarily coded in Solidity

- Extensive developer involvement and supportive community

- Energy costs (or gas fees) associated with transaction handling and executing smart contracts

For organizations eyeing the development of public dapps or delving into **DeFi** arenas, Ethereum emerges as a favored choice.

Hyperledger Fabric

Initiated by the Linux Foundation, Hyperledger Fabric stands out as an open source, permissioned blockchain platform tailored for enterprise scenarios. Some of its notable attributes are as follows:

- An adaptable framework that allows for modifications and personalization

- Multi-language support for smart contracts (known as chaincode)

- Special channels to ensure transaction confidentiality

- Versatile consensus methods

Given its features, Hyperledger Fabric frequently becomes the go-to for businesses keen on integrating blockchain in sectors with regulatory measures, such as banking, healthcare, and logistics.

Corda

Tailored mainly for the world of financial services, Corda is an open source, permissioned blockchain platform. It comes equipped with features such as the following:

- Direct communication pathways among network users

- Backing for intricate financial contracts and smart contracts

- Seamless merging with present financial frameworks and facilities

- Enhanced data privacy and confidentiality tools

Corda, with its specialized attributes, is particularly fitting for financial entities keen on leveraging blockchain for facets such as asset oversight, transaction processing, and other fiscal operations.

Scalability and performance

Scalability and performance are critical considerations when implementing cloud-native blockchain solutions. Organizations must ensure that their chosen platform can handle the anticipated transaction volume and accommodate future growth. Most blockchain networks struggle on both sides in terms of scaling and performing well when the number of transactions grows. Let's examine a couple of strategies that are commonly used in blockchain platforms to achieve better scalability and performance, namely storing data off-chain (that is, not on the blockchain digital ledger), and rolling up multiple blocks and committing them at once. This latter approach is typical in so-called *Layer 2* blockchain networks.

Off-chain storage

Off-chain storage can help improve the scalability and performance of blockchain applications by moving non-essential data and processing off the blockchain. This can help reduce network congestion and transaction fees, while also improving the overall user experience.

Off-chain storage solutions can include the following:

- Traditional databases (for example, Amazon RDS, Azure SQL, and Google Cloud Spanner)

- Distributed databases (for example, Amazon DynamoDB, Azure Cosmos DB, and Google Cloud Firestore)

- File storage services (for example, Amazon S3, Azure Blob Storage, and Google Cloud Storage)

Layer 2 solutions

Layer 2 solutions are built on top of existing blockchain networks and aim to improve scalability and performance by handling transactions and smart contract execution off-chain. Here are some popular Layer 2 solutions:

- **State channels**: Off-chain communication channels that enable participants to transact privately and securely without requiring on-chain transactions

- **Plasma chain**: A framework for creating scalable, hierarchical blockchain networks that rely on the root chain for security

- **Rollups**: Techniques for aggregating and compressing multiple transactions into a single on-chain transaction

State channels

To make this clearer, let's look at examples of each solution. State channels are a Layer 2 scaling solution that allows for off-chain transactions between participants, thereby reducing the load on the main blockchain and increasing transaction throughput. Let's say we have Alice and Bob, who want to engage in multiple transactions with each other without relying on the main blockchain for every transaction. They decide to set up a state channel between themselves to facilitate these transactions. The following figure describes what happens next on a state channel:

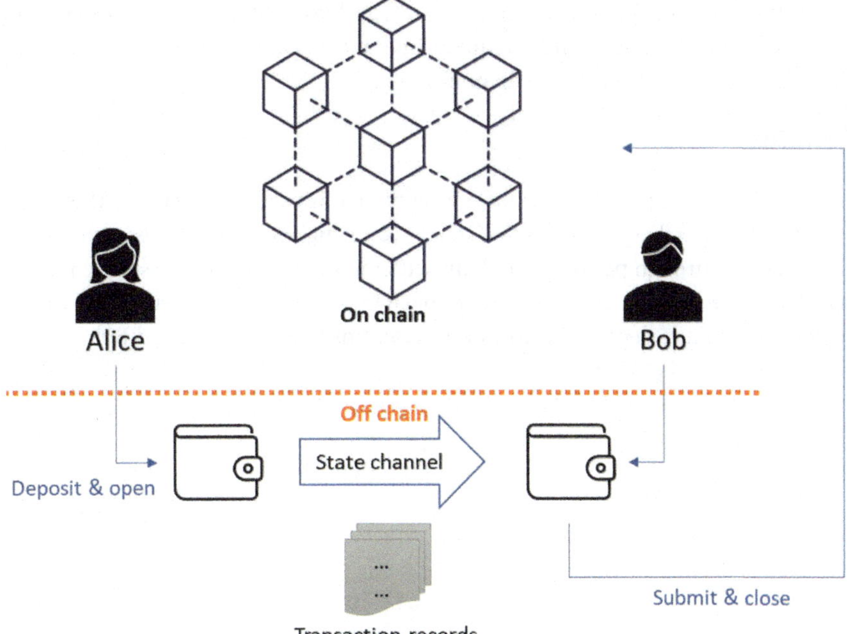

Figure 1.3 – State channel transaction

This is what happens:

1. **Opening the channel**: Alice and Bob create a multi-signature wallet on the blockchain and lock some funds into it as collateral. This collateral serves as security to ensure that both parties abide by the rules of the state channel. The state channel is now open, and Alice and Bob can start transacting off-chain.

2. **Transacting off-chain**: Alice and Bob can now exchange transactions directly with each other off-chain. These transactions are signed by both parties and can involve transferring funds or updating the state of a shared application. Since these transactions are off-chain, they are fast and have minimal fees compared to on-chain transactions.

3. **Updating the state**: As Alice and Bob continue to transact, they keep track of the current state of their interactions. This state includes information such as the balances of each party and any other relevant data. Each time they want to update the state, they exchange and sign a new transaction reflecting the updated state.

4. **Closing the channel**: Once Alice and Bob are done transacting or want to settle their balances on the main blockchain, they can close the state channel. To close the channel, they submit the final state of their interactions to the blockchain. The blockchain verifies the final state and settles any outstanding balances accordingly. The collateral that's locked into the multi-signature wallet is released back to both parties, and the state channel is closed.

By using state channels, Alice and Bob were able to conduct multiple transactions off-chain, reducing congestion on the main blockchain and enjoying fast and low-cost transactions. State channels are particularly useful for scenarios where frequent interactions between parties are needed, such as gaming, microtransactions, or payment channels.

Plasma chains

Moving on to an example of a plasma chain, this approach is used by Layer 2 platforms for creating scalable, hierarchical blockchain networks that can process a high volume of transactions off-chain while maintaining security through periodic on-chain settlement. Let's consider a simple implementation of plasma called *Plasma Cash*, which focuses on token transfers and is often used for NFTs or unique assets. The following figure depicts the flow of transactions between Alice and Bob:

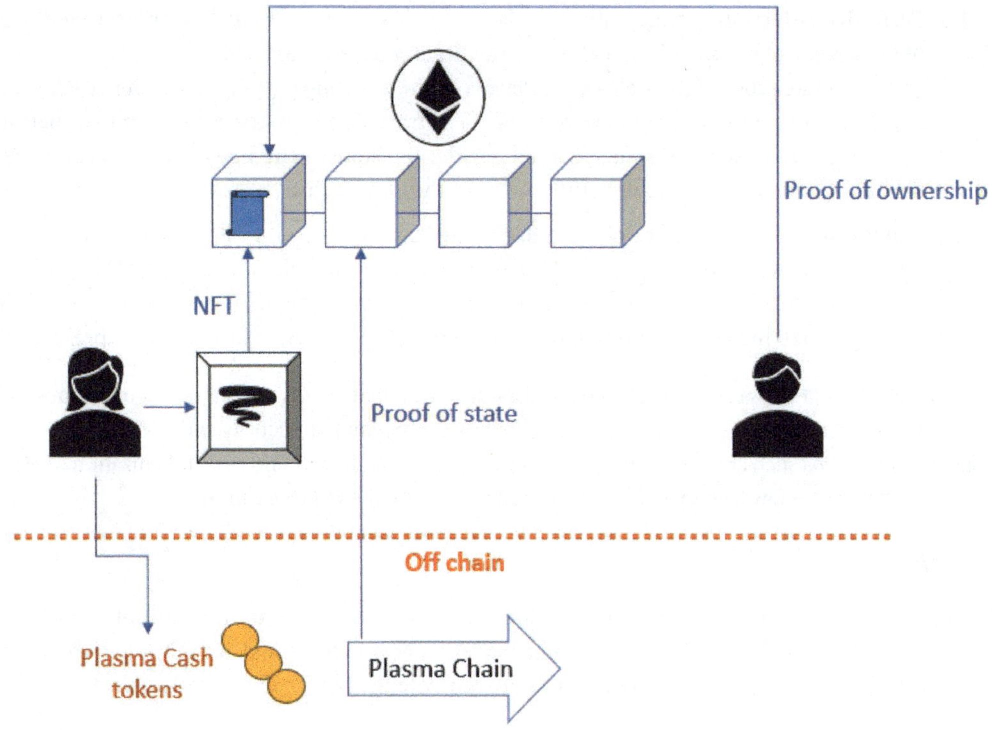

Figure 1.4 – Plasma chain for Layer 2 networks

This is what happens in a plasma chain:

1. **Setup**: Alice wants to trade digital artwork with Bob. She sets up the plasma chain on Ethereum as a smart contract, which acts as the root chain. Alice deposits her digital artwork (represented as an NFT) into the plasma chain's contract, locking it into a specific position in the Merkle tree. This initial state is recorded on the Ethereum mainnet.

2. **Transactions**: Alice and Bob can now trade digital artwork with each other off-chain within the plasma chain. Each transfer of digital artwork is represented by a unique token (for example, Plasma Cash). These transactions are conducted off-chain, allowing for fast and low-cost transfers between Alice and Bob.

3. **Proofs and challenges**: To ensure the security of the plasma chain, participants can challenge invalid transactions by submitting proofs to the root chain (Ethereum). For example, if Bob tries to spend a token that he doesn't own or tries to spend the same token multiple times, Alice or other participants can challenge the transaction by submitting proof of the invalid transaction to the root chain.

4. **Periodic settlement**: Periodically, the plasma chain's operator (Alice) submits a Merkle root of the latest state of the Plasma chain to the Ethereum mainnet. This Merkle root serves as a cryptographic proof of the state of the Plasma chain, allowing participants to verify the validity of transactions without having to process every transaction on the Ethereum mainnet. If no challenges are raised within a specified period, the state of the Plasma chain is considered finalized, and any tokens can be withdrawn from the Plasma chain back to the Ethereum mainnet.

5. **Exit mechanism**: If Alice or Bob want to exit the Plasma chain and withdraw their tokens to the Ethereum mainnet, they submit a proof of ownership (for example, a Merkle proof) to the Plasma chain's contract on Ethereum. The contract verifies the proof and allows the user to withdraw their tokens to the Ethereum mainnet, ensuring that their ownership rights are preserved.

In summary, plasma allows for the creation of scalable blockchain networks by conducting most transactions off-chain, with periodic settlement and on-chain verification to maintain security and trust. This framework enables applications such as decentralized exchanges, gaming platforms, and asset tokenization to achieve high throughput and low latency while leveraging the security of the Ethereum mainnet.

Rollups

Rollups are a common Layer 2 scaling solution that aggregates and submits multiple transactions off-chain to the main blockchain, reducing congestion and increasing throughput. There are two main types of rollups: optimistic rollups and **Zero Knowledge** (**ZK**) rollups. Let's look at an example of how each type works.

Optimistic rollups

Alice wants to participate in a DEX on Ethereum, which is prone to high gas fees and network congestion during peak times. The DEX pictured in the following figure implements an optimistic rollup solution to improve scalability and reduce transaction costs:

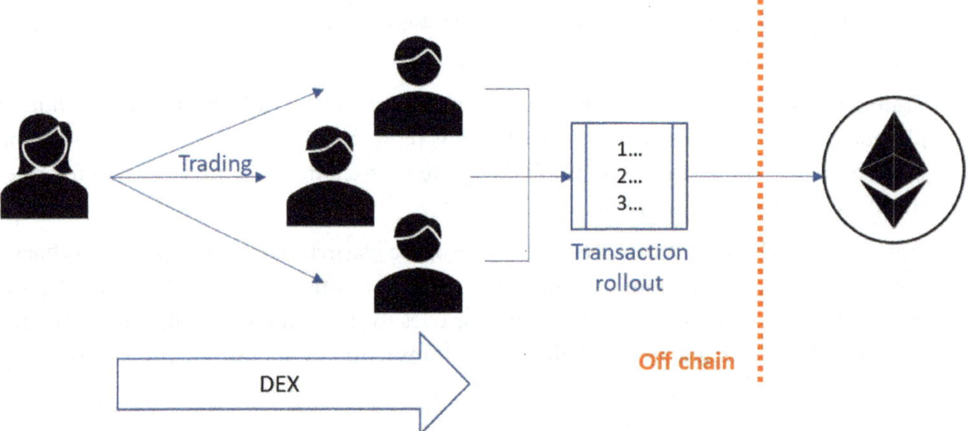

Figure 1.5 – Optimistic rollup transaction process

The transaction process consists of the following steps:

1. **Off-chain transactions**: Alice and other users conduct trades off-chain within the optimistic rollup environment. These transactions are fast and low-cost since they don't require interaction with the Ethereum mainnet. The DEX's optimistic rollup operator aggregates these transactions into a single Merkle root.

2. **Submission to the Ethereum mainnet**: Periodically, the optimistic rollup operator submits the aggregated Merkle root to the Ethereum mainnet, along with a fraud proof that attests to the validity of the transactions. The Ethereum mainnet verifies the validity of the Merkle root and fraud proof. If no fraudulent activity is detected, the transactions are considered valid and included in the Ethereum blockchain.

3. **Challenge period**: After the Merkle root is submitted to the Ethereum mainnet, there is a challenge period during which users can scrutinize the transactions and raise disputes if they suspect any fraudulent activity. If a challenge is raised and proven valid, the fraudulent transactions are reverted, and the guilty party may face penalties.

Optimistic rollups enable high throughput and low-cost transactions by batching multiple transactions off-chain and submitting them to the Ethereum mainnet periodically. Users can enjoy the scalability benefits of Layer 2 solutions while still benefiting from the security and decentralization of the Ethereum mainnet.

ZK rollups

Bob wants to participate in a DeFi protocol on Ethereum, which requires frequent interactions with smart contracts and is affected by high gas fees. The DeFi protocol depicted in the following figure implements a ZK rollup solution to achieve scalability and reduce transaction costs:

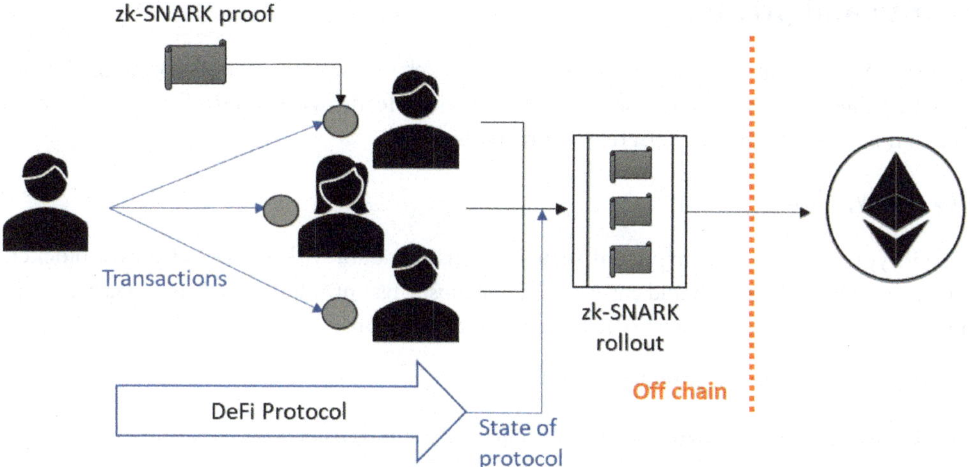

Figure 1.6 – ZK rollup transaction process

This is how the transaction process goes:

1. **Off-chain transactions**: Bob and other users interact with the DeFi protocol off-chain, executing transactions within the ZK rollup environment. These transactions are private and efficient as they don't require interaction with the Ethereum mainnet. The ZK rollup operator generates a succinct proof (zk-SNARK) that attests to the validity of the transactions without revealing sensitive information.

2. **Submission to the Ethereum mainnet**: Periodically, the ZK rollup operator submits the zk-SNARK proof to the Ethereum mainnet, along with a commitment to the updated state of the DeFi protocol. The Ethereum mainnet verifies the validity of the zk-SNARK proof, ensuring that the transactions comply with the protocol's rules without revealing the details of individual transactions.

3. **Finality and settlement**: Once the zk-SNARK proof has been verified, the transactions are considered finalized, and the updated state of the DeFi protocol is reflected on the Ethereum mainnet. Users can interact with the DeFi protocol on-chain, confident that their transactions are secure and valid.

ZK rollups provide high scalability and privacy for blockchain transactions by aggregating off-chain transactions into succinct proofs (zk-SNARKs) that are verified on-chain. Users can enjoy the benefits of efficient and cost-effective transactions while preserving privacy and security.

In summary, both optimistic rollups and ZK rollups are powerful Layer 2 scaling solutions that enable high throughput and low-cost transactions on blockchain networks like Ethereum. They achieve scalability by aggregating and batching transactions off-chain before submitting them to the mainnet, thereby reducing congestion and increasing efficiency.

Security and privacy

Security and privacy are essential aspects of any blockchain implementation. Organizations must ensure that their chosen platform provides the necessary features and controls to protect sensitive data and maintain compliance with relevant regulations.

Data encryption

Data encryption is a crucial aspect of securing sensitive data within a cloud-native blockchain environment. Organizations should ensure that their chosen platform supports the necessary encryption standards and protocols to protect data both at rest and in transit.

Here are some encryption techniques to consider:

- **Transport Layer Security (TLS)** for securing data in transit

- **Advanced Encryption Standard (AES)** for encrypting data at rest

- **Hardware Security Modules (HSMs)** for managing cryptographic keys

Access control and identity management

Access control and identity management are critical for ensuring that only authorized users can access and interact with the blockchain network and its associated resources. Organizations should implement robust access control policies and leverage the identity management features provided by their chosen cloud provider.

Some access control and identity management considerations are as follows:

- **Role-Based Access Control** (**RBAC**) for managing user permissions and privileges
- **Multi-Factor Authentication** (**MFA**) for enhanced user security
- **Single Sign-On** (**SSO**) integration with existing identity providers

It's important to note that access control and identity management are peculiar to permissioned blockchains, where anonymous access is not authorized. **Sovrin** (`https://sovrin.org/`) is a decentralized identity platform that provides tools and protocols for creating and managing self-sovereign identities, where individuals have control over their identity information. Sovrin allows organizations to issue verifiable credentials, such as government-issued IDs or academic certificates, which can be stored and managed by individuals using their Sovrin identities.

Another example is uPort (`https://www.uport.me/`), a decentralized identity platform built on Ethereum. It provides tools and libraries for developers to integrate decentralized identity into their applications, allowing users to control their identity information. Also, uPort supports the creation of self-sovereign identities, enabling users to manage their identity credentials and interact with decentralized applications in a secure and privacy-preserving manner.

Smart contract security

Smart contracts play a crucial role in blockchain applications, and their security is of paramount importance. Organizations should ensure that their smart contracts are thoroughly audited, tested, and secured to prevent vulnerabilities and potential attacks.

Here are some smart contract security best practices:

- Formal verification to prove the correctness of the smart contract code
- Automated testing and fuzzing to identify vulnerabilities
- Security audits conducted by reputable third-party firms

Interoperability and integration

Interoperability and integration are essential considerations for organizations looking to leverage existing systems and infrastructure within their cloud-native blockchain solutions. The chosen platform should support seamless integration with other services and provide the necessary tools and APIs for bridging blockchain networks with traditional systems.

API and SDK support

APIs and SDKs play a crucial role in enabling seamless integration between blockchain networks and existing systems. Organizations should ensure that their chosen platform provides comprehensive API and SDK support for various programming languages and platforms.

The following are some key API and SDK features to consider:

- RESTful APIs for easy integration with web services and applications

- Web3.js, Web3.py, or other SDKs for interacting with Ethereum-based networks

- SDKs for popular programming languages, such as Java, Python, and JavaScript

Blockchain network interoperability

With the expansion and progression of the blockchain universe, the emphasis on ensuring compatibility among diverse blockchain systems has amplified. Organizations need to align with platforms that facilitate communication and interaction across chains, augmenting the effectiveness of their blockchain integrations.

A few notable solutions promoting such inter-chain compatibility are as follows:

- **Cosmos Network**: An integrated network comprising independent, adaptable, and compatible blockchains (`https://cosmos.network/`)

- **Polkadot**: A framework designed to bridge and fortify distinct blockchain entities (`https://polkadot.network/`)

- **Chainlink**: A distributed oracle network that safely links smart contracts to outside data streams and application interfaces (`https://chain.link/`)

Cost optimization

Implementing a cloud-native blockchain solution can be expensive, particularly when considering the costs associated with infrastructure, development, and ongoing maintenance. Organizations should carefully consider the costs involved and optimize their implementations to minimize expenses while maximizing the benefits.

The following are some cost optimization strategies:

- Using managed services to reduce the overhead of managing infrastructure

- Implementing autoscaling policies to ensure efficient resource utilization

- Monitoring and analyzing resource usage to identify and eliminate waste

In this section, we covered the key considerations and best practices for implementing cloud-native blockchain solutions on AWS, Azure, and GCP. By understanding these factors, organizations can make informed decisions and ensure the successful deployment and management of their blockchain applications in a cloud-native environment.

With that, we've discussed the importance of choosing the right blockchain framework, ensuring scalability and performance, securing sensitive data and transactions, enabling seamless interoperability and integration, and optimizing costs. By following these best practices, organizations can successfully implement cloud-native blockchain solutions that meet their specific requirements and deliver the desired benefits.

Summary

In this first chapter, we looked at blockchain and cloud-native technologies, exploring their core characteristics and potential synergies. We touched on the essential features of cloud-native technology, such as containerization and CI/CD pipelines, and the transformative attributes of blockchain, including distributed ledgers and consensus algorithms. We then discussed the merits and trade-offs of melding these technologies for robust, decentralized applications.

Next, we'll start our journey together and delve into practical implementations of cloud-native Web3 applications on platforms such as AWS, Azure, and GCP to create the foundation for crafting innovative and distributed applications.

2

Overview of AWS, Azure, and GCP Services for Blockchain

In this chapter, we will provide an overview of the blockchain services offered by the three major cloud providers: **Amazon Web Services (AWS)**, Azure, and **Google Cloud Platform (GCP)**. We will discuss the various services and tools available for building and deploying blockchain solutions on each platform and cover the strengths, weaknesses, and unique features of each.

Understanding the differences in each cloud offering is critical to building decentralized apps on the most appropriate blockchain infrastructure. In this and the following chapters, we'll dive into the current blockchain services available in the cloud and describe various use cases for them, as well as their potential integration with other cloud services offered by the cloud service provider.

In this chapter, we will cover the following main topics:

- AWS blockchain services
- Azure blockchain services
- GCP blockchain services
- Comparing AWS, Azure, and GCP blockchain services

AWS blockchain services

AWS provides a variety of managed services and tools to help organizations build and deploy blockchain solutions. AWS offers both managed as well as custom deployment options, catering to a wide range of use cases and requirements. In this section, we will introduce the main blockchain services offered by AWS.

Amazon Managed Blockchain

Amazon's Blockchain Management offers a streamlined approach to building and overseeing scalable blockchain networks and supports recognized protocols such as Hyperledger Fabric and Ethereum. It allows organizations to effortlessly establish blockchain networks, launch applications, and oversee network elements.

Let's take a look at the main features of this service:

- Compatibility with Hyperledger Fabric and Ethereum networks

- Hands-off infrastructure management, including node provisioning and network setup

- Collaboration with AWS services, such as Amazon's **Quantum Ledger Database** (**QLDB**) for external data storage and Amazon CloudWatch for tracking

Beginning with *Chapter 4*, we will start working with specific use cases of blockchain in AWS. Specifically, we will discover how to deploy managed blockchain nodes in AWS, how to host a Hyperledger Fabric network on Elastic Kubernetes Service (*Chapter 5*), and how to use Amazon QLDB (*Chapter 6*).

Amazon QLDB

Amazon QLDB is an end-to-end ledger database that ensures a clear, unchangeable, and cryptographically secure transaction record tailored for applications requiring a centralized, reliable authority. It's optimal for transaction scenarios involving various stakeholders, such as financial systems or legal agreements.

Here are the key aspects of Amazon QLDB:

- Automatic scaling due to its serverless framework

- Clear and secure transaction records

- SQL-compatible PartiQL for ledger queries

- Connectivity with AWS tools such as AWS Lambda and Amazon S3 for data management

As mentioned previously, we'll discover QLDB in *Chapter 6*, specifically how to create an instance of the Quantum Ledger, generate a data model, and query data out of it. Please note that the reference to Quantum here is not related to quantum computing. QLDB runs on traditional x64 cloud computing infrastructure.

Amazon EC2 and Amazon Elastic Kubernetes Service (EKS)

Beyond tailored blockchain services, AWS presents the option to design bespoke blockchain networks while utilizing virtual technology and container structures. Amazon's Elastic Computing service EC2 offers adaptable virtual systems for blockchain node hosting, and its Elastic Kubernetes container solution EKS facilitates the deployment of container-based blockchain apps through Kubernetes.

Here are the central features for blockchain deployment using EC2 and EKS:

- Virtual and container setups tailored to blockchain needs

- Adjustability to accommodate fluctuating tasks and network requirements

- Synchronization with AWS tools to enhance storage, tracking, and safety

AWS Marketplace templates

AWS Marketplace provides a variety of templates and solutions from partners that help organizations easily deploy and manage blockchain and distributed ledger technologies. By leveraging these marketplace templates, organizations can quickly set up and configure blockchain networks and applications tailored to their specific needs and requirements.

Here are some of the key features and capabilities of AWS Marketplace templates for blockchain:

- Access to a wide range of partner-provided templates and solutions for implementing blockchain networks and applications

- Support for various blockchain protocols and frameworks, such as Ethereum, Hyperledger Fabric, Corda, and more

- Integration with AWS services for storage, monitoring, and security

Searching for blockchain in the AWS Marketplace generates over 200 entries, as shown in the following screenshot:

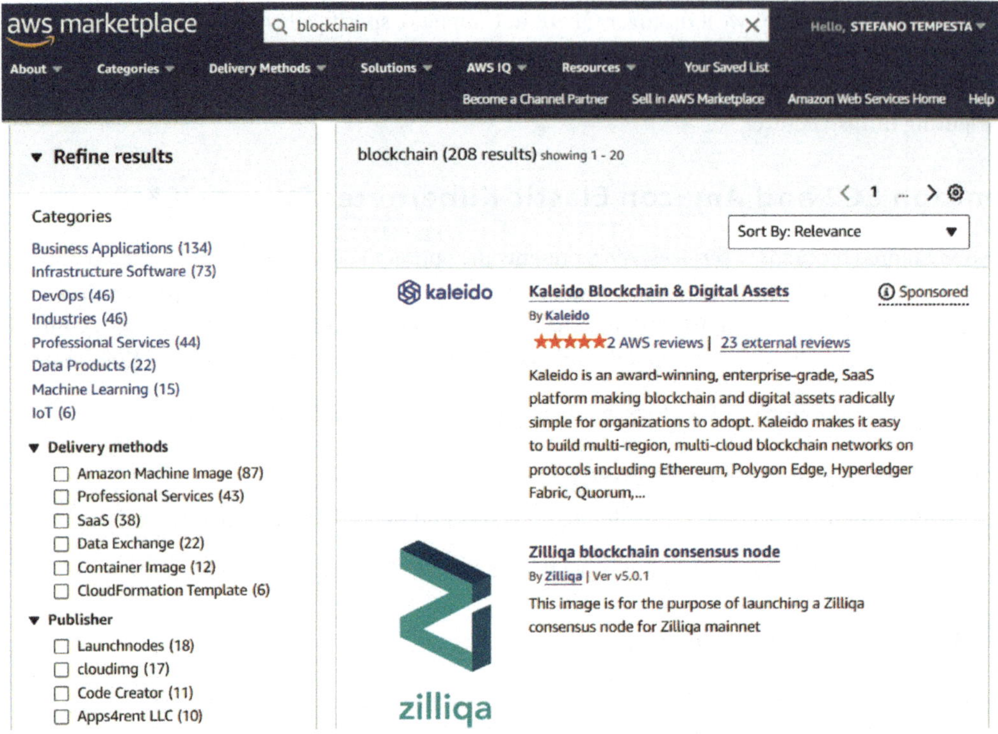

Figure 2.1 – Top search results for blockchain services in AWS Marketplace

This may seem overwhelming, but it demonstrates the rich offerings for blockchain services in AWS. We'll discover a few of them in the next few chapters.

Azure blockchain services

Microsoft Azure offers a variety of services and tools to help organizations build and deploy blockchain solutions. Azure provides managed services, custom deployment options, and partner solutions that cater to a wide range of use cases and requirements. In this section, we will introduce the main blockchain services offered by Azure.

Microsoft Entra Verified ID

Microsoft Entra Verified ID is a decentralized identity platform that enables organizations to create, manage, and verify digital identities on the blockchain. Built on the Sidetree protocol, Entra Verified ID provides secure and privacy-preserving identity solutions.

Here are the key features and capabilities of Microsoft Entra Verified ID:

- You can create and issue verifiable credentials by leveraging prebuilt templates or by specifying rules and design files.

- You can validate a verified ID credential with the user's approval through their digital wallet. This *selective disclosure* approach enables transactions that respect people's privacy by allowing an individual to provide only the necessary pieces of their identity information to a service or a third party, rather than disclosing their full identity record.

- You can revoke or suspend the active verified status of someone's credential.

> **Microsoft Entra Verified ID doesn't store verifiable credential data**
>
> Verifiable credentials are based on a decentralized identity model called **Decentralized Identifiers (DIDs)**. DIDs are owned by the user and stored in their digital wallet. Entra Verified ID enables the issuing and verification of such identifiers (verifiable credentials).

Azure Managed Confidential Consortium Framework (CFF)

Azure Managed CCF is a set of tools and services that enable organizations to create and manage secure, scalable, and decentralized applications using confidential computing technologies. This framework provides the foundation for building consortium networks that require secure data sharing and collaboration among multiple parties.

Here are the key features and capabilities of Azure Managed CCF:

- Support for various blockchain protocols, such as Ethereum and Hyperledger Fabric

- Confidential computing capabilities to protect sensitive data during processing

- Integration with Azure services, such as Azure Key Vault for secure key management and Azure Monitor for logging and monitoring

> **CCF is built on Trusted Execution Environments (TEEs)**
>
> Microsoft CCF, which is built on TEEs, provides a highly secure and scalable platform that's designed to enable confidential computing and data integrity in multi-party computational scenarios, leveraging hardware-encrypted secure memory enclaves to protect data and code from external and internal threats. This technology is at the foundation of confidential computing.

Azure Confidential Ledger

Azure Confidential Ledger offers a fully controlled, unalterable ledger service that's designed to ensure a safe and scalable record for applications that need a centralized, dependable authority. Leveraging confidential computing, this service guarantees the utmost security and privacy of data during its life cycle.

The following are the primary attributes of Azure Secure Ledger:

- An unalterable transaction log enhanced with confidential computing

- Adjustability to accommodate diverse workloads and transaction sizes

- Compatibility with Azure tools such as Azure Identity Management for user verification and Azure Secure Key Storage for key safeguarding

We'll be using Azure Confidential Ledger in *Chapter 9* to store data confidentially, thereby preventing users from accessing such data, even a cloud service provider.

> **Confidential Ledger versus SQL Ledger**
>
> At the time of writing, Microsoft offers two ledger capabilities in Azure: Confidential Ledger and SQL Ledger. The two services are very different since they serve different purposes and are built on different underlying technologies. Azure Confidential Ledger is designed primarily for storing sensitive data in a highly secure, immutable, and tamper-proof environment. It is suitable for scenarios where the integrity and confidentiality of the data are of utmost importance. It is built on the CCF and utilizes TEEs to ensure the security and confidentiality of the data. SQL Ledger, on the other hand, is integrated into Azure SQL Database and is designed for applications that require a tamper-evident ledger database in a familiar SQL environment. It is based on blockchain technology but operates within the SQL Database service, providing blockchain-like functionalities in a relational database context.

Azure Kubernetes Service (AKS)

AKS offers managed orchestration for containers, streamlining the process of launching and overseeing container-based applications. Ideal for blockchain applications, AKS grants organizations the agility and scale for their blockchain structures.

Here are the central features of AKS in a blockchain context:

- Effortless setup and supervision of containerized blockchain solutions

- Adjustability to meet fluctuating tasks and network requirements

- Collaboration with Azure tools to enhance storage, oversight, and protection

We'll be using AKS in *Chapter 7* to deploy a Corda DLT network.

Azure partner solutions

Azure offers various partner solutions for implementing blockchain networks and applications. These solutions include managed services, tools, and resources provided by Azure partners, such as Kaleido Blockchain Service and Calastone DMI Fund Services. By leveraging these partner solutions, organizations can build and deploy blockchain solutions tailored to their specific needs and requirements.

The key features and capabilities of Azure partner solutions for blockchain include access to a variety of managed services, tools, and resources for building and deploying blockchain solutions. Clients also receive direct support from the blockchain provider, rather than generically from Microsoft about Azure, for various blockchain protocols and frameworks. As we'll discover in the next few chapters, integration with other Azure services for storage, monitoring, and security is a key advantage of using blockchain services in Microsoft's cloud:

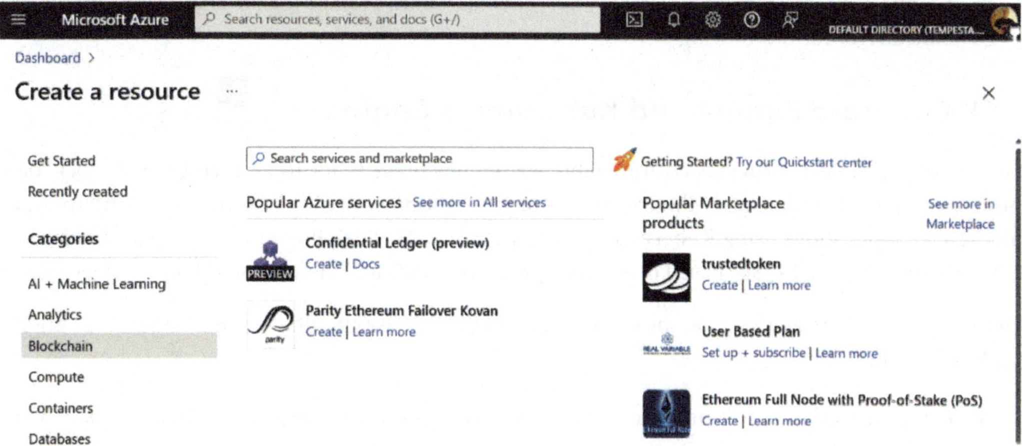

Figure 2.2 – Top list of blockchain services in Azure

Microsoft used to have pioneering blockchain products in Azure, such as Azure Blockchain Service and Azure Blockchain Workbench. Unfortunately, these services have been discontinued and are no longer available in Azure.

GCP blockchain services

GCP offers a variety of services and tools to help organizations build and deploy blockchain solutions. GCP provides the infrastructure and technologies for deploying and managing custom blockchain networks, as well as managed services and partner solutions for implementing a wide range of blockchain use cases. In this section, we will introduce the main services and tools offered by GCP for implementing blockchain solutions.

GCP Blockchain Node Engine

GCP Blockchain Node Engine is a fully managed service that simplifies the deployment and management of blockchain nodes on GCP. The service supports popular blockchain networks such as Ethereum, Bitcoin, and more. With Blockchain Node Engine, users can quickly deploy and scale nodes, ensuring fast and reliable access to blockchain networks.

Here are some of the key features and capabilities of Blockchain Node Engine:

- Support for popular blockchain networks, including Ethereum, Bitcoin, and others
- A fully managed infrastructure that takes care of provisioning nodes, setting up the network, and managing its ongoing operation
- Seamless integration with GCP services, such as Cloud Storage for off-chain data storage and Stackdriver for monitoring

GCP Compute Engine and Kubernetes Engine

GCP Compute Engine provides customizable virtual machines that can be used to host blockchain nodes, while GCP Kubernetes Engine allows organizations to deploy and manage containerized blockchain applications using Kubernetes. These services offer flexibility in deploying and managing custom blockchain networks based on various protocols, such as Ethereum and Hyperledger Fabric.

Here are some of the key features and capabilities of GCP Compute Engine and Kubernetes Engine for blockchain:

- Customizable virtual machine and container configurations to suit the needs of your blockchain network
- Scalability to handle varying workloads and network demands
- Integration with other GCP services for storage, monitoring, and security

We'll be using the blockchain services in GCP, and specifically Kubernetes Engine, in *Chapter 10*.

GCP Marketplace templates

GCP Marketplace provides a variety of templates and solutions from partners that help organizations easily deploy and manage blockchain and distributed ledger technologies. By leveraging these marketplace templates, organizations can quickly set up and configure blockchain networks and applications tailored to their specific needs and requirements.

Let's look at some of the key features and capabilities of GCP Marketplace templates for blockchain:

- Access to a wide range of partner-provided templates and solutions for implementing blockchain networks and applications

- Support for various blockchain protocols and frameworks, such as Ethereum, Hyperledger Fabric, Corda, and more

- Integration with GCP services for storage, monitoring, and security

GCP partner solutions

In addition to Blockchain Node Engine, GCP also offers various partner solutions for implementing blockchain networks and applications. These solutions include managed services, tools, and resources provided by GCP partners, such as ConsenSys, Chainstack, and Digital Asset, among others. By leveraging these partner solutions, organizations can build and deploy blockchain solutions tailored to their specific needs and requirements.

As seen already for the other cloud providers, partner solutions in GCP don't differ much in terms of benefits. Specifically for blockchain services, the key features and capabilities of GCP partner solutions include access to a variety of managed services, tools, and resources for building and deploying blockchain solutions in GCP, direct support for various blockchain protocols and frameworks by the vendor, and strong integration with other GCP services for storage, monitoring, and security:

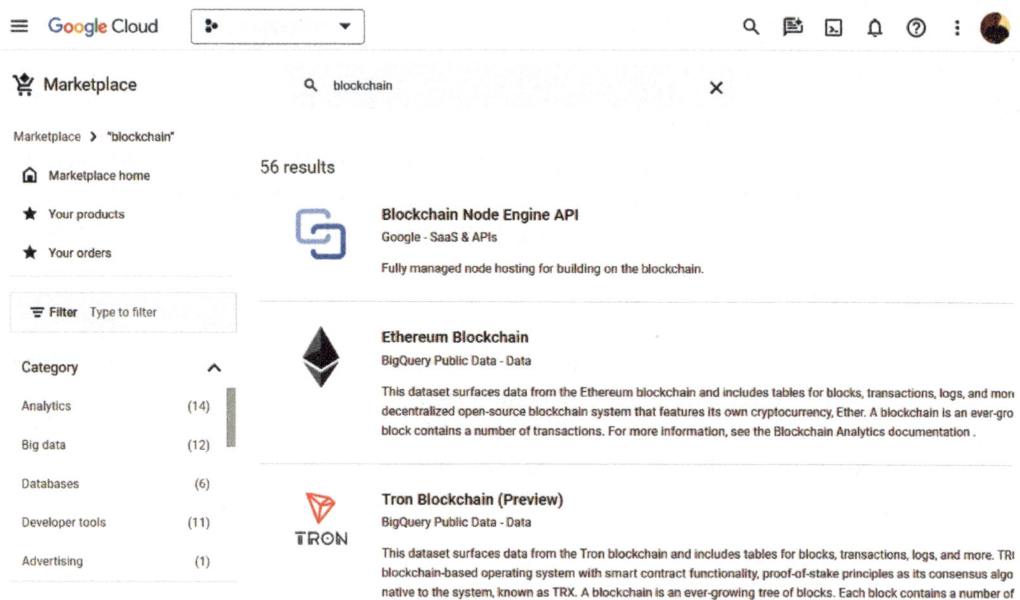

Figure 2.3 – Top list offerings for blockchain services in GCP

As we will learn later in this book, GCP offers a unique proposition for blockchain services that other cloud providers don't offer. The entire Bitcoin and Ethereum datasets are available for exploration with Google BigQuery. All historical data is stored in these datasets and updated every 10 minutes.

Comparing AWS, Azure, and GCP blockchain services

Now that we've introduced the main blockchain services offered by AWS, Azure, and GCP, let's compare their strengths and weaknesses. Each platform has its unique features and capabilities, and choosing the right one depends on your specific needs and requirements.

AWS

AWS offers a comprehensive suite of managed blockchain services, including Amazon Managed Blockchain and Amazon QLDB, as well as support for custom deployments via EC2 and EKS. AWS provides a wide range of features, integrations, and security options, making it a suitable choice for organizations looking for a robust and versatile blockchain platform.

Azure

Azure provides various managed services, custom deployment options, and partner solutions for building and deploying blockchain solutions. With its focus on confidential computing technologies and decentralized identity, Azure is a strong choice for organizations that require secure data sharing and collaboration among multiple parties.

GCP

GCP offers a combination of managed services, such as the Blockchain Node Engine, as well as custom deployment options via Compute Engine and Kubernetes Engine. Additionally, GCP provides a range of partner solutions and marketplace templates, catering to organizations with diverse blockchain requirements.

Strengths and weaknesses

Each cloud provider has its unique strengths and weaknesses. Azure is well-suited for enterprises deeply invested in the Microsoft ecosystem, AWS offers a robust, scalable, and feature-rich environment, and GCP stands out in terms of data analytics and AI/ML integration. The choice between these platforms depends on the specific needs, existing infrastructure, and preferences of the organization.

The following table summarizes the key strong and weak points of each cloud provider, based on the authors' experience in delivering blockchain-based solutions:

Cloud provider	Strengths	Weaknesses
AWS	• **Amazon Managed Blockchain**: Offers a fully managed service that supports popular frameworks such as Hyperledger Fabric and Ethereum • **Scalability**: AWS's infrastructure ensures high scalability and reliability for blockchain applications • **Ecosystem and integration**: Robust integration with AWS's extensive service ecosystem, including AWS Lambda, Amazon S3, and more	• **Cost**: AWS services can be expensive, especially at scale • **Steep learning curve**: The breadth of AWS services can be overwhelming, requiring significant time investment to fully leverage its capabilities
Azure	• **Integration with Microsoft products**: Seamless integration with other Microsoft services and tools such as Azure Active Directory, SQL Database, Power BI, and more • **Enterprise focus**: A strong focus on enterprise needs with robust security features and compliance standards • **Developer tools**: Extensive developer tools and frameworks, such as Azure Blockchain Workbench, which simplifies blockchain app development	• **Limited blockchain protocols**: It may not support as wide a range of blockchain protocols as some competitors • **Complexity**: Some users find the platform complex to navigate and set up, especially for those new to blockchain
GCP	• **Data analytics integration**: Strong integration with GCP's data analytics tools, offering enhanced data analysis capabilities on blockchain data • **AI and machine learning**: Easy integration with Google's AI and machine learning tools, providing advanced analytics capabilities • **Open source-oriented**: Offers strong support for open source technologies, which is beneficial for businesses looking for flexible and customizable solutions	• **Limited dedicated blockchain services**: GCP does not have as many dedicated blockchain services as Azure or AWS • **Focus on partnerships**: GCP often relies on partnerships with blockchain platform providers, which might limit direct control or specific service offerings

Table 2.1 – Comparison of the strengths and weaknesses of cloud blockchain services

Blockchain data security

Blockchain technology inherently emphasizes security, but the way Azure, AWS, and GCP implement and enhance these security features can vary.

Azure is notable for its integration with Microsoft's security services and compliance focus, AWS provides robust managed services and key management options, and GCP offers strengths in data analytics and AI for security, along with a strong commitment to open source technologies. The choice among these platforms should be guided by the specific security needs, compliance requirements, and existing technology stack of the organization.

The following table highlights some key differences in blockchain data security among these three cloud platforms.

Cloud provider	Data security offerings
AWS	• **Amazon Managed Blockchain**: This service automates many aspects of blockchain management, which can enhance security by reducing the risk of human error • **QLDB**: AWS's QLDB offers an immutable and cryptographically verifiable ledger, enhancing data integrity and security • **KMS**: AWS integrates its blockchain services with AWS KMS, which provides strong security for cryptographic keys • **Flexibility in private blockchain networks**: AWS offers flexibility in setting up private blockchain networks, which can be configured for enhanced security based on specific needs
Azure	• **Enterprise integration**: Azure's strong suit is its integration with existing Microsoft security tools and services, such as Azure Entra Verified ID for verifiable identifiers and Azure Key Vault for managing cryptographic keys. • **Compliance and standards**: Azure places a high emphasis on compliance with global and industry-specific standards, which is crucial for enterprises concerned with regulatory requirements • **CFF**: Azure offers unique features such as the CFF, which enhances security and confidentiality in consortium blockchain networks • **Encryption and network security**: Azure ensures data encryption both in transit and at rest, and provides robust network security features

GCP	<ul><li>**Data analytics and AI security**: GCP's strength lies in integrating blockchain data with its powerful data analytics and AI tools, offering advanced security analytics capabilities.</li><li>**Partnerships for blockchain services**: GCP often partners with specialized blockchain providers. The security in this case depends on both Google's cloud infrastructure and the security measures of the chosen blockchain platform.</li><li>**Encryption standards**: Like Azure and AWS, GCP ensures encryption of data in transit and at rest, adhering to strong encryption standards.</li><li>**Open source focus**: GCP's focus on open source technologies means that security can be more transparent and customizable, but this also requires users to be more proactive in managing their security configurations.</li></ul>

Table 2.2 – Key differences in blockchain data security among the three major cloud providers

Before progressing with the next chapters of this book, we recommend that you explore the *Further reading* section at the end of this and each chapter for additional references on the topics presented.

Summary

In this chapter, we delved into blockchain services across AWS, Azure, and GCP, dissecting their distinct features and offerings. Each platform presents a unique blend of managed services, custom deployments, and collaborative solutions, addressing diverse blockchain needs for various organizations.

With this knowledge, we are now equipped to strategize our blockchain endeavors. Whether it's AWS' all-encompassing suite, Azure's emphasis on secure collaborations, or GCP's multifaceted approaches, you can now make informed choices that align with your organizational objectives.

Next, brace yourself for an insightful journey into DevOps practices tailored for cloud-native blockchain. We'll unveil the intricacies of deployment, continuous integration, and seamless delivery, laying the groundwork for you to proficiently implement and oversee blockchain projects on your selected platform.

Further reading

- *Amazon Managed Blockchain Services on AWS*:

  ```
  https://aws.amazon.com/managed-blockchain/
  ```

- *Amazon Quantum Ledger Database*:

  ```
  https://aws.amazon.com/qldb/
  ```

- *Microsoft Web3 Solutions*:

  ```
  https://azure.microsoft.com/solutions/web3/
  ```

- *Microsoft Entra Verified ID*:

  ```
  https://www.microsoft.com/en-au/security/business/identity-
  access/microsoft-entra-verified-id
  ```

- *Azure Managed Confidential Consortium Framework*:

  ```
  https://ccf.microsoft.com/
  ```

- *Google Cloud for Web3*:

  ```
  https://cloud.google.com/web3
  ```

3

DevOps for Cloud-Native Blockchain Solutions

DevOps brings together people, processes, and tools to streamline software development, allowing continuous delivery of value to users, aligned with business goals. As blockchain moves beyond its cryptocurrency roots and into the world of enterprise software, organizations are facing a new challenge: how to apply DevOps principles to this disruptive technology. This chapter delves into the core tenets of DevOps, highlighting best practices and tools that can facilitate continuous delivery, improvement, and infrastructure automation for developing secure and efficient blockchain solutions.

We will cover the following main topics in this chapter:

- Introduction to DevOps for cloud-native blockchain solutions
- CI/CD for blockchain solutions
- IaC for blockchain solutions
- Monitoring and logging for blockchain solutions
- Best practices for DevOps in cloud-native blockchain solutions

Introduction to DevOps for cloud-native blockchain solutions

DevOps is a combination of practices and principles aimed at creating a more efficient and effective collaboration between development and operations teams. By bridging the gap between these teams, organizations can build, test, and deploy software at a faster pace, reduce deployment time, and minimize the risk of errors. The application of DevOps principles in the context of cloud-native blockchain solutions is essential for achieving a streamlined and optimized development life cycle.

In this section, we will learn about the role of DevOps in cloud-native blockchain solutions, and identify the benefits of using DevOps in the development and deployment of blockchain applications.

In particular, we will explore the unique challenges that blockchain solutions present to DevOps practices, such as dealing with **immutable data** structures, **consensus mechanisms**, and the integration of off-chain and on-chain data.

The blockchain challenge for DevOps

Blockchain technology introduces several unique challenges to DevOps practices that necessitate a rethinking of traditional approaches to development, operations, and the **Continuous Integration/ Continuous Deployment (CI/CD)** pipelines. These challenges stem from the inherent characteristics of blockchain, such as immutability, decentralization, consensus mechanisms, and the need for integration between on-chain and off-chain systems. We will discuss these challenges in detail and explore their implications for DevOps.

Dealing with immutable data structures presents the challenge of blockchain's data immutability, which means that once a transaction is added to the blockchain, it cannot be altered or deleted. This immutability is fundamental to the trust and security offered by blockchain but presents challenges in version control, data correction, and updates.

Also, the consensus mechanism poses a challenge: blockchain networks use consensus mechanisms to agree on the validity of transactions. These mechanisms, such as **Proof of Work (PoW)** or **Proof of Stake (PoS)**, have implications for performance, scalability, and security. For example, DevOps must account for the latency and throughput limitations of blockchain networks, which can vary widely based on the consensus mechanism. This may affect how applications are designed, requiring optimizations for transaction processing and state management.

When dealing with the integration of off-chain and on-chain data, **data synchronization and integrity**, as well as **security and privacy**, have implications for DevOps. Ensuring the integrity and consistency of data moving between on-chain and off-chain environments is critical. This requires robust data validation, synchronization mechanisms, and the possible use of oracles (services that feed external data to the blockchain).

> **A word about security and privacy**
>
> The integration points between on-chain and off-chain systems are potential vulnerabilities. DevOps practices must include strong security measures, including encryption, access controls, and regular security audits.

Blockchain technologies necessitate a paradigm shift in DevOps practices, from development and testing to deployment and monitoring. The unique characteristics of blockchain, such as immutability, consensus mechanisms, and the integration of on-chain and off-chain data, introduce challenges that require innovative solutions and a shift towards more rigorous, security-focused development and operational practices. Successful integration of blockchain into DevOps demands a holistic approach, blending blockchain expertise with traditional DevOps capabilities to ensure reliable, efficient, and secure blockchain applications.

Before going any further, let's start by introducing the concept of DevOps and its core principles.

DevOps – Definition and core principles

DevOps is a portmanteau of **development** and **operations**, and it represents a cultural shift that emphasizes collaboration, communication, and integration between these traditionally siloed teams.

The following diagram shows the four pillars of DevOps:

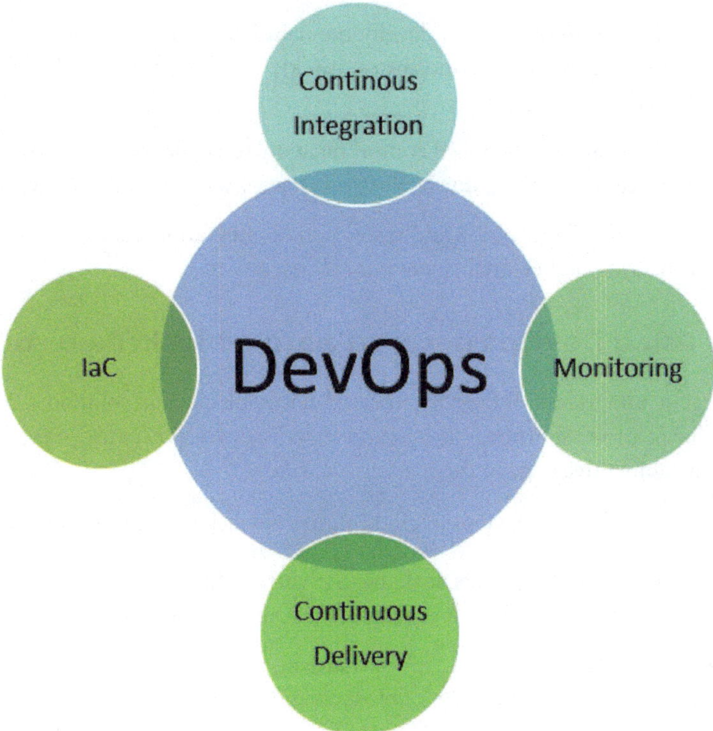

Figure 3.1 – The four pillars of DevOps

The core principles of DevOps include the following:

- **Continuous Integration (CI)**: This involves integrating code changes from multiple developers into a shared repository frequently to identify and fix integration issues early in the development process.

- **Continuous Delivery (CD)**: A method of automatically creating, verifying, and releasing software to production or staging environments with high assurance and dependability.

- **IaC**: This involves managing and provisioning infrastructure through code, which allows for version control, automation, and consistency across environments.

- **Monitoring and logging**: This involves collecting and analyzing data from infrastructure and applications to ensure optimal performance, identify issues, and provide insights for improvement.

The role of DevOps in cloud-native blockchain solutions

Applying DevOps principles to cloud-native blockchain solutions allows for faster development, testing, and deployment of blockchain applications. This is particularly important given the rapid evolution of blockchain technology and the need to quickly adapt to changing requirements and conditions. Some of the key benefits of applying DevOps in cloud-native blockchain solutions include the following:

- **Accelerated time to market**: Faster development and deployment of blockchain applications, enabling organizations to respond more quickly to market demands and stay ahead of the competition.

- **Improved collaboration**: Enhanced communication and collaboration between development and operations teams, breaking down silos and creating a more efficient development process.

- **Reduced risk**: Early identification and resolution of integration issues, minimizing the risk of errors, and ensuring higher-quality software releases.

Benefits of using DevOps in cloud-native blockchain solutions

By understanding the importance of DevOps in cloud-native blockchain solutions, organizations can leverage these principles to create more efficient and effective development processes, leading to faster deployment and a competitive edge in the market.

Some of the key benefits of using DevOps in cloud-native blockchain solutions include the following:

- **Scalability**: The ability to quickly scale blockchain applications and infrastructure to meet changing demands, as well as the ability to scale development and operations processes.

- **Flexibility**: The ability to rapidly adapt to changing requirements and technologies, enabling organizations to stay at the forefront of blockchain innovation.

- **Cost efficiency**: Streamlined development processes and automated infrastructure management can lead to reduced costs associated with both development and operations.

- **Security**: Improved collaboration between development and operations teams can lead to more secure and compliant blockchain applications.

Let me provide a few examples so you can better understand the impact of DevOps on blockchain solutions:

Imagine a **supply chain management** system built on a private blockchain such as Hyperledger Fabric. This platform would track the movement of goods from origin to destination, ensuring transparency and efficiency. Initially, the network may only handle a low volume of transactions. When demand increases, manually scaling the infrastructure (servers, storage, etc.) becomes a complex and time-consuming task. IT teams need to provision new resources, configure them, and integrate them with the existing system. This can lead to downtime and delays. During periods of low activity, underutilized resources result in wasted costs. Manually scaling down can be equally challenging, leaving the system vulnerable to overprovisioning. Also, the solution experiences limited agility; responding to sudden changes in demand becomes difficult, as the lack of automation hinders the ability to quickly adapt to market fluctuations or unexpected events.

> **Note**
>
> The problem with the scalability of public blockchain networks such as Bitcoin or Ethereum is that adding multiple nodes to the network doesn't increase the number of **Transactions Per Second (TPS)** the system can process. Due to the decentralized nature of the network, only specific mining or staking nodes get the authority to process the transaction, while other nodes wait for the block creation. The only possible ways to increase the TPS of a blockchain is to roll up transactions (an approach taken by Layer 2 networks) or create sidechains that can grow independently. More information is available in this article: `https://shardeum.org/blog/what-is-blockchain-scalability/`.

Now, let's see how having fully automated DevOps practices can help improve CI/CD. Changes to the chaincode are automatically tested and deployed, improving the speed and reliability of updates. This allows for faster scaling of the network to meet changing demand. In Hyperledger Fabric, for example, we can upgrade a chaincode using the same Fabric life cycle process that we initially used to install and start the chaincode. This includes the upgrade of the chaincode binaries as well as the update of any relevant chaincode definition policies. More information on the life cycle of chaincode in Hyperledger Fabric can be found here: `https://hyperledger-fabric.readthedocs.io/en/release-2.5/chaincode_lifecycle.html`.

Also, the infrastructure is defined as code, enabling automated provisioning and configuration of resources. As demand increases, new resources are automatically spun up in the cloud, seamlessly scaling the platform. This is called IaC. Lastly, real-time monitoring of the blockchain network and its resources allows for proactive identification and resolution of bottlenecks. This ensures optimal performance and scalability.

In this scenario, the supply chain system will reduce downtime, as automated infrastructure provisioning and configuration minimize the risk of human errors and downtime during scaling operations. Overall, all parties involved experience increased operational efficiency; automating routine tasks frees up IT teams to focus on more strategic initiatives.

Imagine now a blockchain platform used for **identity management** in a healthcare system. This platform stores and verifies sensitive patient data, making security paramount. Without DevOps in place, security vulnerabilities are often identified late in the development life cycle, requiring costly rework and potential data breaches. Moreover, applying security patches across the distributed blockchain network can be slow and prone to human error, leaving the system exposed for longer periods. The lack of real-time monitoring and logging makes it difficult to detect and respond to security incidents quickly.

Let's apply the principles of DevOps to this solution, starting with security considerations that are integrated into the CI/CD pipeline from the beginning, preventing vulnerabilities from being introduced in the first place. This practice is normally called **security by design** static code analysis, vulnerability scanning, and penetration testing become part of the automated build and deployment process. Security patches are automatically deployed across the blockchain network, ensuring consistent and timely updates on all nodes. This minimizes the window of vulnerability and reduces the risk of attacks. Real-time monitoring and logging tools provide visibility into the network's security posture, allowing for early detection and rapid response to security incidents. Specifically to the blockchain component of the solution, the inherent immutability of the digital ledger makes it tamper proof, ensuring data integrity and preventing unauthorized modifications.

In addition to these benefits, the healthcare institute will also experience improved compliance, as DevOps practices help organizations meet security regulations and standards more effectively, as well as increased trust and transparency; a secure blockchain platform fosters trust and transparency in data management, especially valuable in healthcare applications.

These are just a few examples of how blockchain applications can benefit from good DevOps practices. In the next section, we'll look in more detail at the common patterns and technologies for delivering effective and robust CI/CD in blockchain.

CI/CD for blockchain solutions

CI and CD are essential DevOps practices that can be implemented for cloud-native blockchain solutions. CI/CD helps automate the creation, verification, and release processes, resulting in faster and more dependable software deliveries.

In this section, we will define CI and CD in the context of blockchain solutions, understand a few popular CI/CD tools used for blockchain development, and discuss the implementation of CI/CD for blockchain applications.

CI and CD

Together, CI/CD form a pipeline that streamlines software development, testing, and deployment of blockchain solutions. These practices are fundamental in achieving agile and efficient software development workflows, enabling teams to deliver high-quality software with speed and reliability.

CI involves frequent integration of code changes from multiple developers into a shared repository. The main goal of CI is to integrate changes made to the software by different developers in the team into a central repository frequently, typically several times a day. Whenever a developer commits changes to the source code, automated build and test processes are triggered. This ensures that the new code integrates well with the existing code and does not introduce bugs.

CD extends the concept of CI by ensuring that, in addition to integration, the new code is always in a deployable state. After the CI process of building and testing, CD involves automated steps to deploy the changes to a staging or production environment. This makes it possible to release new updates to customers quickly and efficiently.

In the context of blockchain solutions, CI/CD plays a crucial role in enabling the rapid development, testing, and deployment of blockchain applications and smart contracts. However, automation cannot perform miracles if the design of smart contracts is wrong from the beginning. This is why it's important to consider a correct approach to coding smart contracts first, and then apply automation in testing and deployment.

Smart contract design

Blockchain apps handle sensitive tasks such as finances, critical business processes, and member/customer confidentiality. Errors in these areas could seriously harm your company or consortium. So, blockchain apps need stricter risk management and testing than usual software. A common approach is to treat smart contract design as microservice design: break down the solution into core entities and their processes, then build separate, modular smart contracts for each. This lets them evolve independently over time.

Common practices for writing robust smart contracts include the following:

- Make your code fail quickly and clearly, preventing unexpected behavior.

- Always check conditions first, then modify the contract state, and finally interact with other contracts. This minimizes risk and improves readability.

- Whenever possible, let users initiate payments (**pull**) instead of pushing funds directly (**push**). This avoids unexpected code execution and potential errors.

- Break down your code into clear, independent modules. This makes it easier to understand, test, and maintain, ultimately leading to more secure smart contracts.

- Although time consuming, writing comprehensive tests is crucial for preventing future issues and ensuring the long-term stability of your smart contracts.

Let's take the supply chain example given before and pretend we're writing a contract that, given the name of a party in the supply chain managed on blockchain, returns its state. Try and spot a few code smells in the following code snippet:

```solidity
contract SupplyChain {
    uint constant DEFAULT_STATE = 1;
    mapping(string => uint) partyStates;

    function getPartyState(string memory name) public view returns
(uint) {
        if (bytes(name).length != 0 && partyStates[name] != 0) {
            return partyStates[name];
        }
        else {
            return DEFAULT_STATE;
        }
    }
}
```

The `getPartyState` function might seem flawless, returning the party's state as expected. But it harbors a hidden danger. Instead of throwing an error when invalid input is provided, it silently returns a default value. This seemingly harmless behavior can be a recipe for disaster. To give a metaphor for real life (I define *real life* as anything that is not coding 😊), think of a faulty smoke detector. Instead of blaring an alarm when it detects smoke, it simply flashes a harmless green light. This false sense of security could lead to delayed action and potentially devastating consequences. Similarly, hiding errors in this function can mask real problems, leaving the supply chain system vulnerable to unforeseen failures.

A more correct approach should consider checks for conditions before returning the desired value, and also the separation of precondition checks, so as to make each condition fail separately with a proper failure message:

```solidity
function getPartyState(string memory name) public view returns (uint)
{
    if (bytes(name).length == 0) revert("Invalid name.");
    if (partyStates[name] == 0) revert("Name not found.");

    return partyStates[name];
}
```

There are several good design patterns to keep in mind when creating a smart contract. This was just an example to consider because it defines how test scripts would be produced and then automated as we introduce CI/CD in our blockchain software development life cycle. Test scripts require test data. However, data on a blockchain is immutable and permanently persisted. There's no wipeout of dummy data after a test battery is executed. How can we automate test data generation for blockchain solutions? Let's have a look at a few common approaches in the next section.

These code snippets are available in the GitHub repository: `https://github.com/PacktPub-lishing/Developing-Blockchain-Solutions-in-the-Cloud`

Smart contract testing

Testing smart contracts is crucial due to their immutable nature and the significant financial and operational implications of bugs or vulnerabilities. A comprehensive testing strategy for smart contracts typically encompasses several layers, including unit testing, integration testing, and, for highly critical contracts, formal verification. Each of these testing layers targets different aspects of the smart contract's reliability and security.

Unit testing involves testing the smallest parts of a smart contract in isolation, typically individual functions or methods, to ensure that they perform as expected under various conditions. Strategies include the following:

- **Mocking external calls**: Use mocking frameworks to simulate external calls and dependencies, allowing developers to test the contract's logic in isolation

- **Parameterized testing**: Employ parameterized tests to cover a wide range of inputs, including edge cases, to ensure functions handle all expected and unexpected inputs correctly

- **Automated test suites**: Utilize automated testing frameworks (e.g., Truffle, Hardhat for Ethereum, etc.) to write and run unit tests, ensuring they are executed regularly

Integration testing assesses the interactions between different parts of the smart contract system, including interactions between multiple contracts and the interaction with the blockchain itself. Strategies include the following:

- **Testing contract interactions**: Ensure that contracts work together as expected, simulating real-world scenarios where contracts call each other or pass data

- **Simulating blockchain conditions**: Use blockchain simulation tools to test contracts in an environment that closely mirrors live blockchain conditions, including gas usage and transaction ordering

- **Cross-contract dependencies**: Test how your contract responds to changes in contracts it depends on, if applicable, ensuring that your system is resilient to changes in external contracts

In the blockchain space, there's a process known as **formal verification** that utilizes mathematical methods to prove the correctness of smart contracts or protocols. It involves using formal methods, such as mathematical logic and automated reasoning techniques, to verify that a smart contract or protocol behaves as intended under all possible scenarios. Formal verification is particularly useful for mission-critical contracts where the cost of failure is high. Strategies include the following:

- **Modeling contract logic**: Translate smart contract logic into a formal model that can be analyzed mathematically. This often involves using specialized languages and tools.

- **Property specification**: Clearly define the properties and invariants a contract must maintain (e.g., no unauthorized access, conservation of value, etc.) and use formal verification tools to prove these properties are always upheld.

- **Tooling**: Leverage formal verification tools and platforms (e.g., K Framework, CertiK, etc.) designed for smart contracts.

A comprehensive testing strategy for smart contracts is essential for ensuring their reliability and security. By combining unit tests, integration tests, and formal verification methods, developers can address different aspects of contract functionality and security. Integrating these testing practices into a continuous development and deployment pipeline, complemented by peer reviews and professional audits, maximizes the reliability and security of smart contracts in the blockchain ecosystem.

Test data generation

While the foundation of DevOps remains the same, blockchain's distributed nature and immutable ledger introduce specific considerations. These require careful assessment when tailoring a DevOps approach for blockchain solutions. As mentioned, a typical challenge with testing smart contracts is the generation of sample data. Filling test data in a blockchain is trickier than other databases. In a regular database, you can use scripts to create tables and fill them with data. But in a blockchain, it's not so straightforward. Adding new blocks is complex, as it involves validation by all network members. While scripting can work, it requires careful coordination and sequencing, especially when multiple parties are involved. A common approach for testing purposes is to fork an existing blockchain, basically making a copy, and using that for testing. This can save us the hassle of making fake transactions from scratch.

Imagine a blockchain as a highway. Every car on the road is a transaction. Forking is like building a detour, a new path for the cars to take. Sometimes, the detour is temporary, like for road repairs. This is a **soft fork**. Other times, the detour becomes the new highway, and the old one is abandoned. This is a **hard fork**. In both cases, the cars (transactions) have to choose which path to follow. The following figure shows how a hard fork deviates from the original blockchain, creating a new chain of its own:

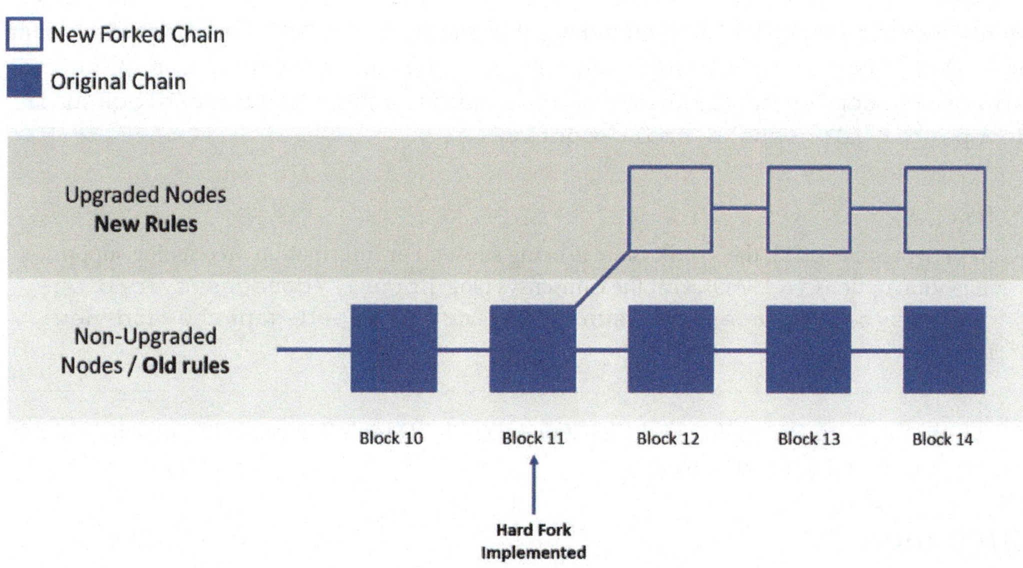

Figure 3.2 – Hard fork of a blockchain

When a soft fork happens, the blockchain protocol keeps the rules defined on the network backward compatible. Blocks in the forked chain will follow the old rules as well as the new rules. Blocks in the original chain will continue to follow the old rules. The following figure depicts the alternate chain, which rejoins the original one when a consensus is reached:

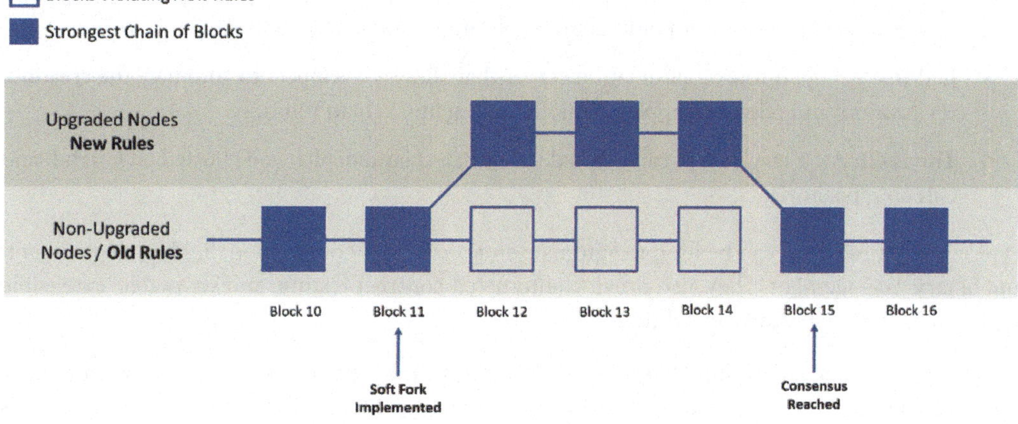

Figure 3.3 – Soft work of a blockchain

More information about hard and soft forking as a strategy for test data generation can be found in the *DevOps for Blockchain Smart Contracts* article written by the author: `https://learn.microsoft.com/en-us/archive/msdn-magazine/2019/october/blockchain-devops-for-blockchain-smart-contracts`

> **Note**
>
> As of September 2023, the Truffle Suite is being sunset. For information on ongoing support, migration options, and FAQs, visit the Consensys blog: `https://consensys.io/blog/consensys-announces-the-sunset-of-truffle-and-ganache-and-new-hardhat`.

It's now time to learn about a few of the most popular CI/CD tools that can help automate and thus improve our blockchain development.

CI/CD tools

There are several popular CI/CD tools that can be used for blockchain development, including the following:

- **GitHub Actions**: A workflow automation tool that allows developers to build, test, and deploy code directly from GitHub repositories. GitHub Actions can be used to automate CI/CD pipelines for blockchain applications and smart contracts.

- **GitLab CI/CD**: A CI/CD tool integrated with GitLab, providing a seamless CI/CD experience for developers working with blockchain applications and smart contracts.

- **Jenkins**: An open source automation server that allows developers to automate the creation, verification, and release of applications, including blockchain solutions.

- **Travis CI**: A CI service that can be used to build, test, and deploy code hosted on GitHub and GitLab repositories.

Typically, all these tools can be used to automate smart contract compilation, linking, deployment, and binary management. They also provide automated contract testing, and scriptable, extensible deployment, and migration capabilities.

They are not specific to handling blockchain networks, though. For example, we may want to have automated network management for the deployment of nodes. More specific blockchain CI/CD tools include the following:

- **Truffle Suite**: This has already been mentioned, and is unfortunately being sunset. Available as open source on GitHub at `https://github.com/trufflesuite`.

- **Hardhat Network**: A comprehensive testing framework for Ethereum with features such as mocking blockchain behavior and deploying contracts to testnets.

- **Embark**: A development framework for Ethereum offering tools for building, testing, and deploying smart contracts, including a built-in CI/CD pipeline.

- **DAppChain**: A platform specifically designed for CI/CD in blockchain development, providing tools for automated testing, deployment, and monitoring.

- **Blockpress**: A cloud-based platform offering CI/CD services specifically for Hyperledger Fabric applications, including smart contract testing and deployment.

Let's take Hardhat, one of the most popular tools, for reference. Setting up Hardhat for automating the deployment of smart contracts involves several steps, which include the following:

1. Install Node.js (version 14 or above) and npm.

2. Create a Hardhat project with the following bash command. Follow the prompts to complete the setup:

   ```
   npx hardhat
   ```

3. **Configure Hardhat**: This involves specifying networks, compilers, and other settings in the `hardhat.config.js` file similar to the following example:

   ```
   require("@nomiclabs/hardhat-ethers");
   module.exports = {
     solidity: "0.8.4",
     networks: {
       rinkeby: {
         url: "https://rinkeby.infura.io/v3/your_project_id",
         accounts: ["0xYOUR_PRIVATE_KEY"],
       },
     },
   };
   ```

   Replace `your_project_id` and `0xYOUR_PRIVATE_KEY` with your Infura project ID and private key respectively.

4. Now, we'll look at automating the deployment of a smart contract with Hardhat Network, which requires the definition of the pipeline in a JavaScript file. Let's call this file `deploy.js`. The script first defines the contract factory by indicating the name of the smart contract, using the `hardhat` library. Then, it deploys the contract by calling the `deploy()` method on the contract factory:

   ```
   const hre = require('hardhat');

   async function main() {
     const contractFactory = await hre.ethers.
   getContractFactory('Contract Name');
     const myContract = await contractFactory.deploy();
   ```

```
    await myContract.deployed();

    console.log('Contract deployed to:', myContract.address);
}

main()
  .then(() => process.exit(0))
  .catch((error) => {
    console.error(error);
    process.exit(1);
  });
```

5. To run the script, we just need to use the `npx hardhat` bash command with the indication of the deploy file:

```
npx hardhat deploy deploy.js
```

Similar scripts can be prepared for the other tools. This is an example of contract deployment automation, but what about the required infrastructure to run the contract? We also want to automate the deployment of IaC, as described in the next section.

IaC for blockchain solutions

IaC is a DevOps practice that involves configuring and deploying infrastructure through code. This approach allows for version control, automation, and consistency across environments, making it easier to scale and manage blockchain applications and infrastructure. In this section, we will explore popular IaC tools for managing blockchain infrastructure, and discuss the implementation of IaC for blockchain solutions. Let's start by understanding the benefits of IaC.

IaC – Definition and benefits

IaC is the method of configuring and deploying infrastructure through code, rather than manual processes or proprietary tools. The consistent use of IaC eventually leads to a more streamlined and efficient management of the required infrastructure to run a software solution. It's easy to identify a few benefits of IaC:

- **Faster deployment**: IaC enables the rapid provisioning and configuration of infrastructure, reducing the time it takes to deploy blockchain applications and infrastructure

- **Scalability**: IaC makes it easier to scale blockchain applications and infrastructure to meet changing demands, ensuring optimal performance and resource utilization

- **Consistency**: By using code to manage infrastructure, organizations can ensure consistency across environments, reducing the risk of errors and inconsistencies that can lead to downtime or security vulnerabilities

- **Version control**: IaC allows for version control of infrastructure configurations, making it easier to track changes, roll back to previous configurations, and collaborate on infrastructure management

Popular IaC tools for blockchain infrastructure management

There are several popular IaC tools that can be used for managing blockchain infrastructure, including the following:

- **Terraform**: An open source IaC tool that enables developers to define and manage IaC using a declarative configuration language. Terraform supports multiple cloud providers, including AWS, Azure, and **Google Cloud Platform (GCP)**, making it a versatile choice for managing blockchain infrastructure.

- **AWS CloudFormation**: A service provided by AWS that allows developers to define, manage, and update infrastructure resources using JSON or YAML templates. CloudFormation is specifically designed for AWS environments and can be used to manage blockchain infrastructure on the AWS platform.

- **Azure Resource Manager (ARM) templates**: A JSON-based template language used to define and manage Azure resources. ARM templates enable developers to manage blockchain infrastructure on the Azure platform using a declarative syntax.

- **Google Cloud Deployment Manager**: A service provided by GCP that allows developers to define and manage infrastructure resources using YAML templates. Deployment Manager is specifically designed for GCP environments and can be used to manage blockchain infrastructure on the GCP platform.

To be clear, these tools are IaC automation tools for any type of software application, not only blockchain-based solutions. There are a few other specific tools that are meant for managing blockchain infrastructure only:

- **Hyperledger Fabric Composer**: Designed specifically for Hyperledger Fabric, Composer offers a visual interface and templates for managing networks, channels, and smart contracts

- **DAppChain**: A cloud-based platform with built-in IaC features for deploying and managing blockchain applications across various platforms

- **Embark**: A popular Ethereum development framework with tools for managing network configurations, deploying smart contracts, and automating infrastructure provisioning

- **Blockpress**: A cloud-based platform specifically for Hyperledger Fabric, offering advanced IaC features for network deployment, resource management, and scaling

- **Constellation Network**: A blockchain platform with its own built-in IaC features for managing nodes, networks, and smart contracts

Choosing the right tool depends on a number of factors, not least the specific blockchain platform of reference. Some tools are platform specific, while others offer broader compatibility. Some tools also require more programming knowledge than others, or require more custom scripting, as opposed to a simpler visual configuration. Not least, budget and licensing preferences are a factor of consideration too. Some tools are open source and free, while others offer paid plans with additional features.

Implementing IaC for blockchain solutions

We will have plenty of script examples in the next chapters of this book, but just to give a quick example now, let's look at how to use Embark to automatically deploy a Hyperledger Fabric network with two nodes (one orderer and one peer) on a cloud platform such as GCP.

The following steps assume that you have access to GCP and are familiar with the GCP console, and the services offered. If this is not the case yet, we will talk about deploying and implementing blockchain solutions on GCP in *Part 4* of the book, so feel free to skip this example for now and return to it later after completing *Part 4* of the book.

The required steps for configuration are as follows:

1. **Define the network configuration**: First, we create a directory for storing a configuration file with the network settings (e.g., `network/config.json`). Within `config.json`, we specify details such as the number of orderer and peer nodes, their resource requirements, and the desired deployment environment (GCP in this case):

    -   `cloud`: This specifies the cloud platform being used (GCP in this case)

    -   `gcpConfig`: This provides specific configurations for GCP, such as the project ID and the zone where to deploy the nodes as well as the type and operating system of the virtual machine for the nodes

    -   `nodes`: This defines the number of nodes to provision for each type

    The following example can be used as a template for the `config.json` file to describe the initial resources needed in the configuration of the blockchain network in the GCP:

```
{
    "cloud": "gcp",
    "gcpConfig": {
        "projectId": "YOUR_GCP_PROJECT_ID",
        "zone": "YOUR_GCP_ZONE",
        "machineType": "e2-micro",
        "imageProject": "debian-cloud/debian-11"
    },
    "nodes": {
        "orderer": {
            "count": 1,
```

```
        "services": ["orderer"]
      },
      "peer": {
        "count": 1,
        "services": ["peer"]
      }
    }
  }
```

2. **Write the deployment script**: Create a JavaScript script (e.g., `deploy.js`) that uses the Embark API to interact with the network configuration and deploy your smart contracts.

   The script will load the network configuration from `config.json`, and use Embark's `deployNode` function to provision and configure the orderer and peer nodes on GCP. This function will deploy the smart contracts to the network using Embark's deployment functionalities, and perform any additional tasks such as setting up channels or fetching contract addresses:

```javascript
const Embark = require('embarkjs');
const config = require('./network/config.json');

(async () => {
  try {
    const embark = new Embark(config);
    await embark.deployNodes({
      type: 'node',
      count: 1, // One peer node
      cloud: 'gcp',
      gcpConfig: {
        // GCP project ID and other relevant details
      },
    });
    await embark.deployNodes({
      type: 'orderer',
      count: 1, // One orderer node
      cloud: 'gcp',
      gcpConfig: {
        // GCP project ID and other relevant details
      },
    });

    // Wait for nodes to be provisioned
    await embark.plugins.blockchain.ensureNetworkReady();

    // Deploy smart contracts
    await embark.contracts.deploy('MyContract');
```

```
    // Get the deployed contract address
    const contractAddress = embark.contracts.MyContract.address;

  } catch (error) {
    console.error(error);
    process.exit(1);
  } finally {
    // Clean up resources
    await embark.plugins.blockchain.cleanNetwork();
  }
})();
```

3. Integrate with CI/CD pipelines, for example, using a tool such as Hardhat Network, or similar. We have seen the configuration of Hardhat previously in this chapter, and how to prepare a deployment script. This allows for automated deployment of the network and smart contracts on every code push or specific trigger.

In the next section, we'll look into best practices and tools for improving the quality of blockchain solutions by constantly monitoring the usage of the network.

Monitoring and logging for blockchain solutions

Monitoring and logging are essential aspects of managing cloud-native blockchain solutions. Collecting and analyzing data from blockchain nodes and infrastructure allows organizations to ensure optimal performance, identify issues, and provide insights for improvement. In this section, we will explore popular monitoring and logging tools for blockchain infrastructure and applications, and understand how to implement monitoring and logging practices for blockchain solutions. Let's start by looking at the importance of monitoring and logging for blockchain solutions.

Importance of monitoring and logging for blockchain solutions

Monitoring and logging are critical components of managing and maintaining cloud-native blockchain solutions. Effective monitoring and logging practices provide several benefits, including the following:

- **Performance optimization**: By continuously monitoring and analyzing data from blockchain nodes and infrastructure, organizations can identify performance bottlenecks and optimize their applications and infrastructure for better performance

- **Issue detection and resolution**: Monitoring and logging enable organizations to quickly identify and resolve issues, minimizing downtime and ensuring the reliability of their blockchain solutions

- **Security and compliance**: Monitoring and logging help organizations detect security vulnerabilities, track access and usage patterns, and maintain compliance with industry regulations and standards

- **Insights for improvement**: Analyzing data from monitoring and logging systems can provide valuable insights into how to improve the efficiency, scalability, and security of blockchain solutions

Popular monitoring and logging tools for blockchain solutions

Several popular monitoring and logging tools can be used to collect and analyze data from blockchain nodes and infrastructure, including the following:

- **Native logging solutions**: Cloud providers such as AWS, Azure, and GCP offer native logging and monitoring solutions for collecting and analyzing data from infrastructure and applications. These solutions include Amazon CloudWatch, Azure Monitor, and Google Cloud's operations suite (formerly Stackdriver).

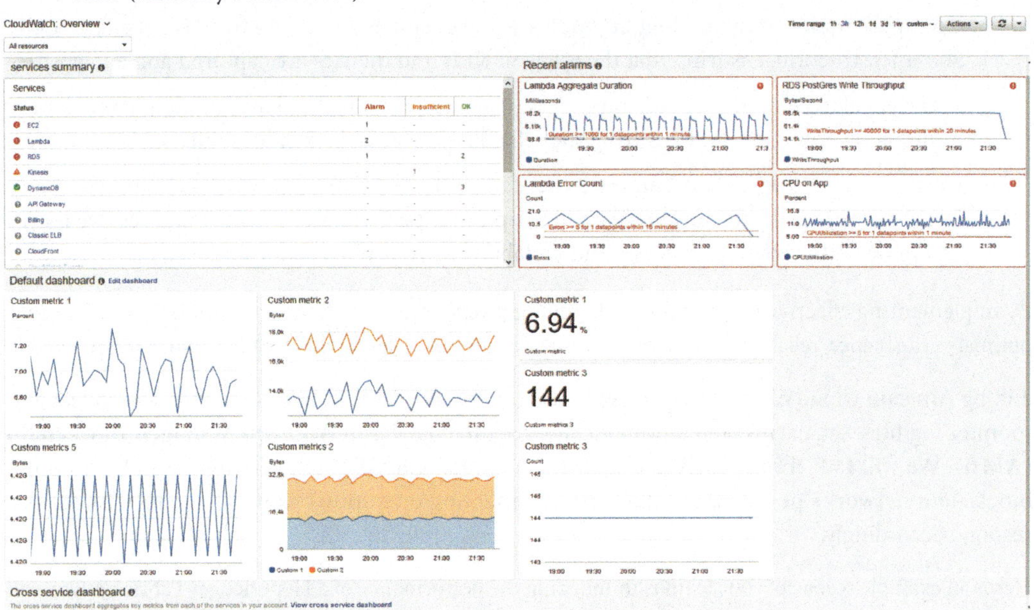

Figure 3.4 – A dashboard in Amazon CloudWatch visualizing different metrics

- **Prometheus**: An open source monitoring and alerting toolkit designed for reliability and scalability. Prometheus can be used to collect metrics from blockchain nodes and infrastructure, providing real-time insights into the performance and health of your blockchain solution.

- **Grafana**: An open source visualization and analytics platform that can be used to create interactive dashboards for monitoring and analyzing data from blockchain nodes and infrastructure. Grafana integrates with various data sources, including Prometheus, Elasticsearch, and native cloud provider monitoring solutions.

- **Elasticsearch, Logstash, and Kibana (ELK stack)**: A popular open source log management and analysis platform that can be used to collect, process, and visualize data from blockchain nodes and infrastructure. The ELK stack provides a powerful solution for monitoring and analyzing logs and metrics from blockchain applications.

Implementing monitoring and logging for blockchain solutions

To implement effective monitoring and logging practices for blockchain solutions, the following steps can be followed:

1. Identify the **Key Performance Indicators** (**KPIs**) and metrics that are most relevant to a blockchain solution, such as transaction throughput, latency, node health, and resource utilization.

2. Choose a monitoring and logging tool or platform that best fits business and technical needs, and integrates with the chosen cloud provider or platform.

3. Configure monitoring and logging tools to collect and analyze data from blockchain nodes and infrastructure, ensuring that the relevant KPIs and metrics are captured and analyzed.

4. Establish alerting and notification mechanisms to proactively identify and address issues, minimizing downtime and ensuring the reliability of your blockchain solution.

5. Continuously monitor and analyze the data collected from your blockchain nodes and infrastructure, using the insights gained to optimize performance, identify and resolve issues, and improve the overall efficiency and scalability of your blockchain solution.

By implementing effective monitoring and logging practices for blockchain solutions, we can ensure optimal performance, reliability, and security while gaining valuable insights for continuous improvement.

Taking Amazon CloudWatch as an example, we can use it to collect and track metrics, collect and monitor log files, set alarms, and automatically react to changes in **Amazon Managed Blockchain** (**AMB**). We will talk about AMB in detail in *Chapter 4*. CloudWatch provides insights into the blockchain network's performance and activity, enabling operations teams to identify issues and respond accordingly.

Here's an example scenario: your team is monitoring the performance of a Hyperledger Fabric network on AMB. You want to ensure that transaction processing times remain low and that the network is healthy.

You can set up CloudWatch to monitor metrics such as `BlockHeight`, `SuccessfulTransactionsPerSecond`, and `FailedTransactionsPerSecond`. By tracking these metrics over time, you can identify trends or spikes that may indicate problems.

If you notice a sudden increase in `FailedTransactionsPerSecond`, you can set up an alarm in CloudWatch to notify your operations team. The team can then investigate the issue, which might involve checking for smart contract errors, network configuration issues, or resource constraints.

You may also be concerned about the resources your AMB network is consuming, particularly CPU and memory utilization, as they directly impact the cost and performance of your blockchain network.

You can set up alarms in CloudWatch for when CPU utilization or memory usage goes above a certain threshold for an extended period. These thresholds are determined based on historical usage patterns and performance requirements.

Once an alarm is triggered, automated actions can be initiated, such as sending a notification to an SNS topic that alerts the operations team. The team might respond by scaling the network resources, optimizing chaincode, or investigating inefficient transactions.

These examples illustrate how Amazon CloudWatch, when utilized effectively, can enhance the monitoring and management of AMB networks, ensuring they run smoothly, efficiently, and cost effectively.

Best practices for DevOps in cloud-native blockchain solutions

There is no technology that can work automatically and magically without a proper culture and the adoption of best practices. Organizations can improve collaboration between development and operations teams, streamline development processes, and ensure a more efficient and effective deployment of blockchain applications, by following a few recommendations:

- **Foster a culture of collaboration**: Encourage open communication and collaboration between development and operations teams, breaking down silos and promoting a shared responsibility for the success of your blockchain solutions.

- **Automate as much as possible**: Implement automation across the development life cycle, including CI, CD, infrastructure provisioning, and monitoring. Automation helps to reduce manual tasks, minimize the risk of human error, and ensure a more efficient and reliable development process.

- **Implement IaC**: Use IaC tools to manage and provision infrastructure, ensuring consistency, scalability, and version control across environments. IaC can help streamline infrastructure management, reduce the risk of errors, and enable more efficient deployment of blockchain solutions.

- **Leverage monitoring and logging tools**: Utilize monitoring and logging tools to collect and analyze data from your blockchain nodes and infrastructure. This will help to optimize performance, identify and resolve issues, and provide insights for continuous improvement.

- **Establish a branching strategy**: Define a branching strategy for your code base that promotes frequent integration of code changes and ensures that your CI/CD pipeline is triggered for every code change. This helps to identify and fix integration issues early in the development process, ensuring higher-quality software releases.

- **Continuously improve**: Encourage a culture of continuous learning and improvement, regularly reviewing and refining your DevOps practices to ensure they remain effective and relevant as your blockchain solutions evolve and grow.

- **Test rigorously**: Implement thorough testing practices, including unit testing, integration testing, and performance testing, to ensure the quality and reliability of your blockchain applications and smart contracts.

- **Prioritize security**: Ensure that security is considered at every stage of the development process, incorporating security best practices and tools to safeguard your blockchain applications and infrastructure from potential threats.

- **Keep documentation up to date**: Maintain comprehensive and up-to-date documentation for your blockchain applications, infrastructure, and DevOps processes. This will help to ensure that all team members have a clear understanding of your solution and can collaborate more effectively.

If we were to identify priorities in the adoption of DevOps best practices, the following three core areas would define the implementation process as depicted in the following image:

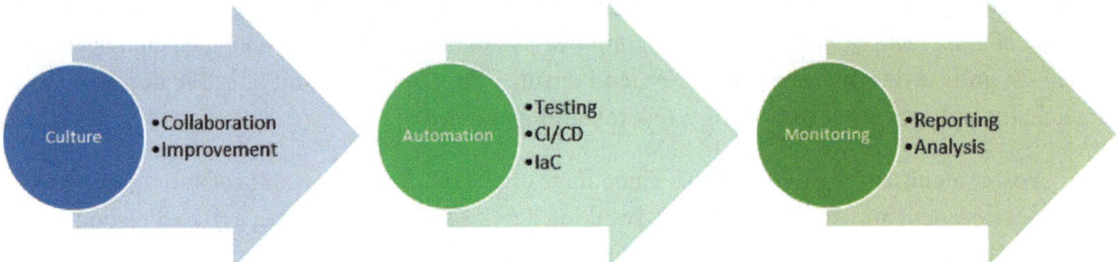

Figure 3.5 – Best practices for DevOps

By following these best practices for DevOps in cloud-native blockchain solutions, organizations can create a more collaborative environment, streamline development processes, and ensure more efficient and effective deployment of blockchain applications.

Summary

This chapter showed how DevOps can enhance cloud-native blockchain solutions by speeding up and ensuring development and deployment. We covered the benefits of introducing a robust DevOps practice for the development of cloud-native blockchain solutions and described how DevOps can lower deployment time and error risk, and improve the scalability, business continuity, and security of a blockchain solution. We then went into the details of CI/CD and IaC tools for blockchain, which is how to automate the build, test, and deployment of smart contracts, as well as how to configure and deploy blockchain infrastructure. The final section focused on aspects of monitoring and logging, with specific attention to blockchain nodes.

This chapter completes the first part of the book, dedicated to providing an introduction to cloud-native blockchain. In *Part 2*, we'll be looking at the technologies, best practices, and tools for deploying and implementing blockchain solutions on AWS.

Part 2: Deploying and Implementing Blockchain Solutions on AWS

This part covers deploying and implementing blockchain solutions on AWS, including how to host a blockchain network on Elastic Kubernetes Service, getting started with Amazon Managed Blockchain, and using Amazon Quantum Ledger Database.

This part includes the following chapters:

- *Chapter 4, Getting Started with Amazon Managed Blockchain*
- *Chapter 5, Hosting a Blockchain Network on Elastic Kubernetes Service*
- *Chapter 6, Building Records with Amazon Quantum Ledger Database*

4

Getting Started with Amazon Managed Blockchain

Building and managing a blockchain network can be challenging as it requires significant expertise and resources to set up, configure, and maintain the network infrastructure and software. To address these challenges, AWS offers **Amazon Managed Blockchain (AMB)**, a fully managed service that enables you to easily create and manage scalable blockchain networks using popular open source frameworks.

We will cover the following main topics in this chapter:

- Introduction to AMB
- Creating a managed blockchain network
- Inviting members and managing access
- Deploying and managing nodes
- Key considerations for security, scaling, and monitoring
- Building a tracking application

Technical requirements

To run the scripts presented in this chapter for provisioning a managed blockchain service on AWS, you will need an AWS account and access to the AWS management console at `https://console.aws.amazon.com`. If you do not have an AWS account, you can create one by following the instructions on the AWS signup page: `https://repost.aws/knowledge-center/create-and-activate-aws-account`.

All scripts are available in this book's GitHub repository at `https://github.com/PacktPublishing/Developing-Blockchain-Solutions-in-the-Cloud`.

Introduction to AMB

AMB is a fully managed service that enables us to easily create and manage scalable blockchain networks using popular open source frameworks such as Hyperledger Fabric and Ethereum. It streamlines the process of setting up, managing, and scaling a blockchain network, allowing us to focus on building our applications instead of the underlying infrastructure.

In this section, we will introduce the features and benefits of AMB, providing a foundation for understanding how it can be used to build blockchain applications on AWS.

The features and benefits are illustrated in the following figure, and described in the following two subsections:

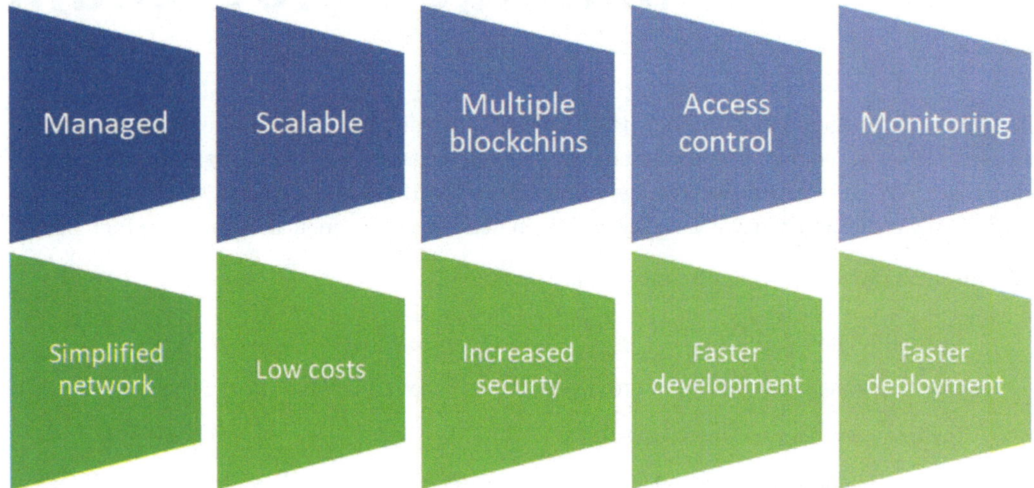

Figure 4.1 – The features and benefits of AMB

Features of AMB

AMB provides several key features that make it an attractive choice for deploying blockchain solutions:

- **Fully managed service**: AMB takes care of the operational aspects of running a blockchain network, such as provisioning hardware, configuring software, and managing network resources. This allows us to focus on developing our applications without worrying about the underlying infrastructure.

- **Scalable and reliable**: The service is built on AWS infrastructure, which ensures high availability, fault tolerance, and scalability. We can easily add or remove nodes to adjust the network capacity as needed, and AMB will automatically replicate data across multiple availability zones for redundancy and fault tolerance.

- **Multiple blockchain frameworks**: AMB supports popular blockchain frameworks, such as Hyperledger Fabric and Ethereum, enabling us to select the framework that best suits our needs.

- **Fine-grained access control**: We can control access to our blockchain network by inviting members and setting up permissions for them. This allows us to create a secure and private environment for our blockchain applications.

- **Monitoring and logging**: AMB integrates with Amazon CloudWatch and AWS CloudTrail, enabling us to monitor our blockchain network and access logs for troubleshooting and auditing purposes.

Benefits of AMB

Using AMB for our blockchain applications offers several benefits:

- **Simplified network management**: The fully managed nature of the service allows us to focus on building your applications while AMB takes care of the operational aspects of the network.

- **Lower costs**: By leveraging the AWS infrastructure, we can benefit from economies of scale and pay only for the resources we use. This can result in lower costs compared to running a blockchain network on-premises or on virtual machines.

- **Increased security**: AMB provides built-in security features, such as encryption at rest and in transit, and integrates with other AWS services for additional security, such as AWS **Key Management Service (KMS)** for key management and AWS **Identity and Access Management (IAM)** for access control.

- **Faster development and deployment**: With AMB, we can easily set up a blockchain network and start building our applications without the need for extensive knowledge of blockchain infrastructure or network management.

Now that we have an understanding of the features and benefits of AMB, let's move on to creating a managed blockchain network on AWS.

Choosing a blockchain framework

AMB supports two popular blockchain frameworks: Hyperledger Fabric and Ethereum. Choosing between them depends on the specific requirements of the application, the business context, and the desired features. The following table provides a comparison between the two frameworks to help you determine which might be more suitable for a given use case when operating within AWS:

	Hyperledger Fabric	**Ethereum**
Type of network	Hyperledger Fabric is a permissioned network, meaning that participants are known and verified, which is often a requirement for enterprise applications that demand privacy and security.	Ethereum provides the flexibility to build applications on a public network (main net) or a private network. This is suitable for **decentralized applications** (**dapps**) that might benefit from broader network effects.
Transaction and token handling	Hyperledger Fabric supports private transactions and private data collections, which enable transactions that are confidential between involved parties.	Ethereum supports a variety of token standards, such as ERC-20 and ERC-721, which are widely used for creating fungible and **Non-Fungible Tokens** (**NFTs**), respectively.
Consensus mechanism	Fabric has a modular and customizable architecture that allows for plug-and-play components in consensus and membership services. The consensus mechanism in Fabric is not about mining but rather about reaching an agreement on the order of transactions and ensuring consistency of the ledger.	Ethereum's main net has used Proof of Stake since the release of Ethereum 2.0. Private Ethereum networks on AWS can use alternative consensus mechanisms such as **Proof of Authority** (**PoA**), which can be more scalable and less resource-intensive.
Programmability	Fabric uses chain code (smart contracts) to encapsulate business logic, and it can be written in general-purpose programming languages such as Go, Java, and Node.js.	**Ethereum Virtual Machine** (**EVM**) is a powerful component of Ethereum, allowing developers to write smart contracts in languages such as Solidity or Viper.

Performance	Generally, it offers higher transaction throughput and faster consensus algorithms, which are more suitable for enterprise applications.	When operating a private Ethereum network on AMB, performance can be significantly improved by using alternative consensus protocols such as PoA, which are more suited to permissioned environments and can provide faster and more predictable transaction validation times.
Use cases	Fabric is best for B2B environments such as supply chain, finance, and healthcare where data privacy and permissioned access are a priority.	Ethereum is ideal for applications that require a decentralized network, such as **Decentralized Finance (DeFi)**, open marketplaces, or where tokenization is necessary.

Table 4.1 – Comparison between Hyperledger Fabric and Ethereum in AMB

When you're deciding on one framework over the other, go through a list of requirements for the solution being developed. For example, you might need to choose between **decentralization** and **centralization**. If the application requires or benefits from a decentralized ecosystem with many anonymous participants, Ethereum might be the best choice. For applications that require strict privacy and known identities, Hyperledger Fabric is preferable.

How about the **consensus** mechanisms? Ethereum's mainnet consensus mechanism (Proof of Stake) is different from Fabric's more flexible and less resource-intensive options, such as Raft or Kafka.

On the **development ecosystem** side, Ethereum has a vast array of tools, wallets, and dapps already available, which can be advantageous if interoperability with existing systems is a requirement.

On the converse, when talking of **compliance** and data regulation, Hyperledger Fabric is often better suited for applications with stringent data governance and compliance requirements due to its permissioned nature.

Does the application need to issue tokens? If the application revolves around the creation and management of tokens, Ethereum's native support for tokens via ERC standards makes it a strong candidate.

In conclusion, both Ethereum and Hyperledger Fabric have distinct advantages and are supported by AMB, which simplifies the setup and management of these networks. The choice between the two should be guided by the specific requirements of the project, including factors such as the need for privacy, the type of consensus mechanism, and the existing developer ecosystem.

Creating a managed blockchain network

In this section, we will cover the steps involved in creating a managed blockchain network on AWS. We will discuss how to configure the network for optimal performance and choose the appropriate consensus algorithm for our use case.

To get started with AMB, sign into the AWS management console (`https://aws.amazon.com/console/`). Search for `managed blockchain`; you will see the **Amazon Managed Blockchain** service at the top of the results, as shown in the following screenshot:

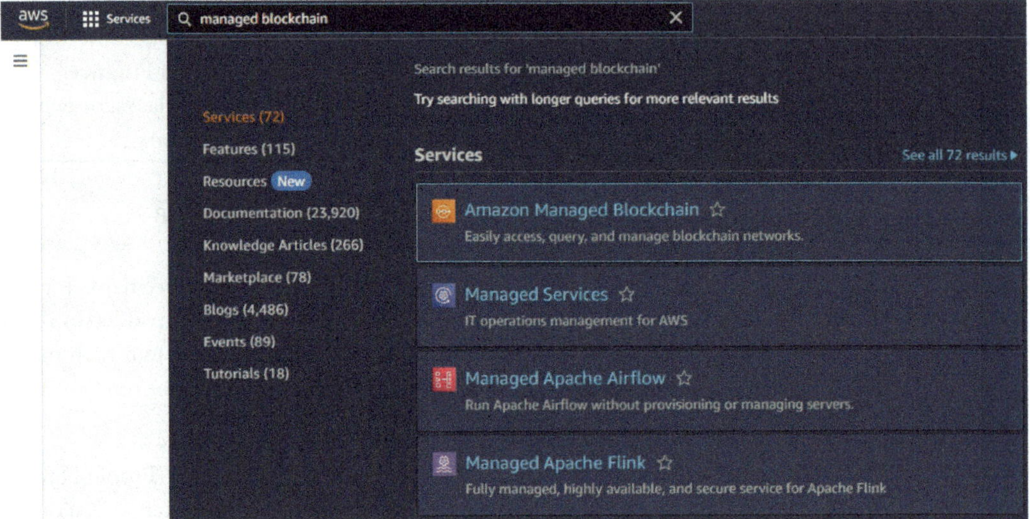

Figure 4.2 – Search results in the AWS console for managed blockchain

Now, follow these steps to provision the AMB service:

1. From the list of services in the search results, select **Amazon Managed Blockchain**; this will open the configuration page for AMB.

2. Click on the **Private networks** tab and then the **Create private network** button in the AMB console:

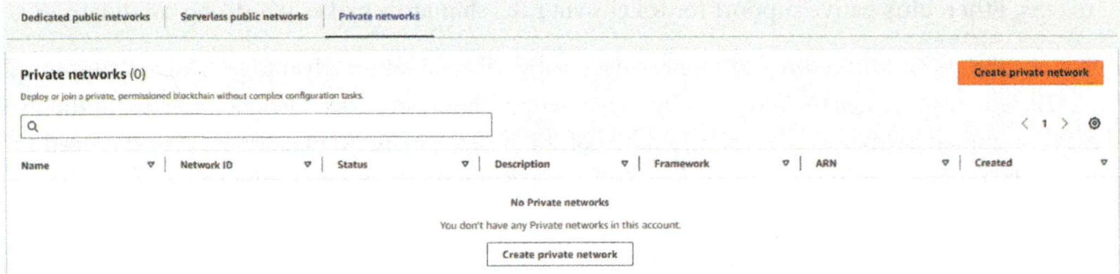

Figure 4.3 – Creating a private network in the AMB console

3. Choose the blockchain framework you want to use for your network. We can choose between **Hyperledger Fabric** and **Ethereum**. For this example, we will choose **Hyperledger Fabric**:

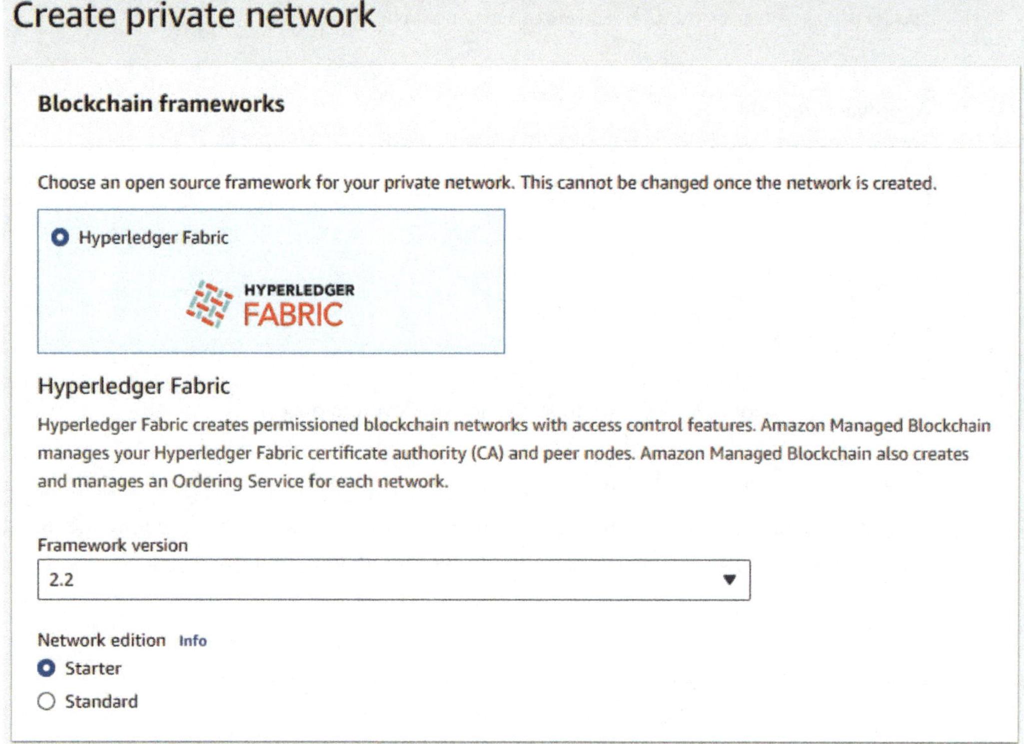

Figure 4.4 – Selecting Hyperledger Fabric as our blockchain framework

Configure the network settings:

- Enter a network name that is unique and easily identifiable.

- Choose a network description to provide additional context about the network's purpose.

- Select the **Hyperledger Fabric** edition. We can choose between the **Starter** edition, which is suitable for development and testing, and the **Standard** edition, which is designed for production environments.

4. Configure the network's voting policy. This policy determines how decisions are made within the network, such as adding or removing members. We can choose between the following options:

- **Simple majority**: A proposal is approved if more than 50% of the members vote in favor

- **Super majority**: A proposal is approved if more than a specified percentage of the members vote in favor, such as 80% or 90%

We can specify the required percentage when configuring the policy:

Voting policy Info

Specify the percentage of Yes votes required to approve a proposal.

Approval threshold

Specify the percentage of Yes votes required to approve a proposal.

| Greater than ▼ | 50 | % |

Proposal duration

Specify how long proposals are open for voting in 1-hour increments up to 168 hours maximum.

| 24 | hour(s) |

Figure 4.5 – Configuring the network's voting policy

5.   Click on **Next** to proceed to the member configuration process.

6.   Enter a unique name for the founding member of the network. This member will have the initial administrative privileges and be able to invite other members to join the network:

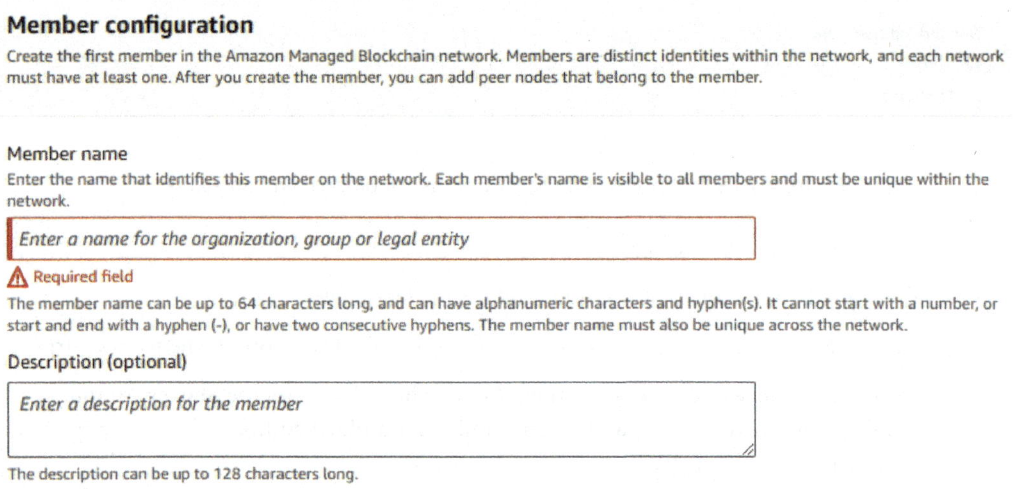

Member configuration

Create the first member in the Amazon Managed Blockchain network. Members are distinct identities within the network, and each network must have at least one. After you create the member, you can add peer nodes that belong to the member.

Member name

Enter the name that identifies this member on the network. Each member's name is visible to all members and must be unique within the network.

| Enter a name for the organization, group or legal entity |

⚠ Required field

The member name can be up to 64 characters long, and can have alphanumeric characters and hyphen(s). It cannot start with a number, or start and end with a hyphen (-), or have two consecutive hyphens. The member name must also be unique across the network.

Description (optional)

| Enter a description for the member |

The description can be up to 128 characters long.

Figure 4.6 – Member configuration for joining an existing AMB network

7.   Click on **Next** to review the network configuration. It is still possible to make any necessary changes by clicking on the **Edit** button next to each section.

8.   Click on **Create network and member** to start the process of creating our AMB network.

The network creation process can take some time, typically around 30 minutes. We can monitor its progress in the AMB console. Once the network has been created, we can proceed with inviting members and managing access.

Inviting members and managing access

In this section, we will cover how to invite members to the blockchain network and manage their access to the network. We will discuss setting up permissions for members and managing their access to the network.

Inviting members to the network

To invite other members to join the network that we have previously deployed, we just need to follow these steps:

1. In the AMB console, navigate to the **Networks** tab and select your network.

2. Click on the **Members** tab and then click on **Propose invitation** to create an invitation proposal:

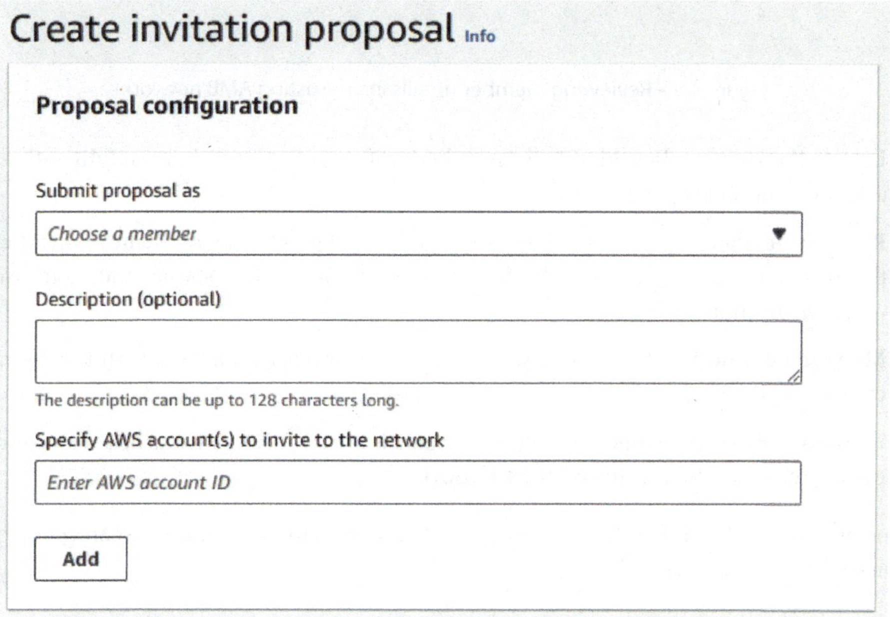

Figure 4.7 – Inviting members to join an existing AMB network

3. Enter the AWS account ID of the member you want to invite. It is possible to invite multiple members at once by entering their account IDs on multiple lines.

4. Click **Create** to send the invitation to the specified AWS account(s).

The invited member(s) will receive an invitation in their AMB console. They will need to accept the invitation to join the network.

Managing member access and permissions

Once a member has joined the network, we can manage their access and permissions. In the AMB console, navigate to the **Members** tab and select the member you want to manage.

We can perform the following actions:

- **View member details**: Review the member's information, such as their AWS account ID, voting policy, and network permissions:

Details		
Member ID m-67C57IKB35ELVC4WAW2FLREADI	Created Tue, Jan 16, 2024	Fabric certificate authority endpoint Info ca.m-67c57ikb35elvc4waw2flreadi.n-se4d74f6v5aqtiqa2bkj2fm4by.mana gedblockchain.us-east-1.amazonaws.com:30002
Description -	ARN arn:aws:managedblockchain:us-east-1:243458509621:members/m-67C57 IKB35ELVC4WAW2FLREADI	KMS Key AWS_OWNED_KMS_KEY
Status ⊘ Available		

Figure 4.8 – Reviewing member details in an existing AMB network

- **Update the voting policy**: Modify the member's voting policy by clicking on **Edit voting policy** and choosing a new policy.

- **Remove member**: Remove a member from the network by clicking on **Remove member**. Note that this action is irreversible and will permanently remove the member and their associated resources from the network.

- **Manage peer nodes**: Add, modify, or remove peer nodes associated with the member by clicking on the **Peer nodes** tab.

- **Create and manage channels**: Create new channels for the member to participate in, or manage existing channels by clicking on the **Channels** tab.

With members invited and their access managed, let's move on to deploying and managing nodes on the managed blockchain network.

Deploying and managing nodes

In this section, we will cover how to deploy and manage nodes on the managed blockchain network. We will discuss adding and removing nodes and monitoring the health of the network.

Adding a new node

To create a new node for a member in the deployed network, follow these steps:

1. In the AMB console, navigate to the **Members** tab and select the member for whom we want to create a new node.

2. Click on the **Peer nodes** tab and then click **Create node**:

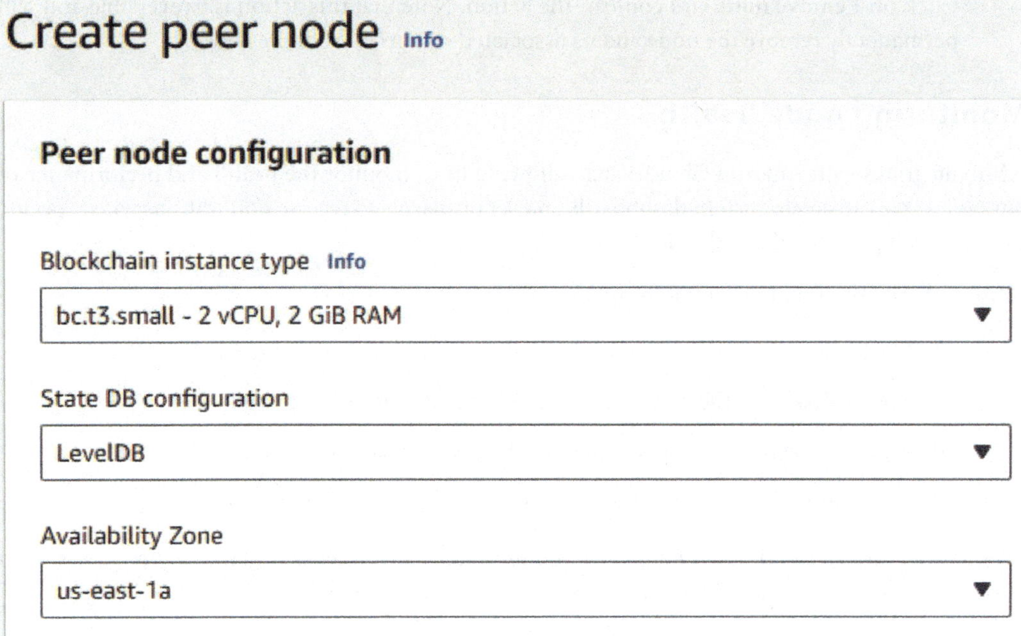

Figure 4.9 – Adding a new node for an existing member

Configure the new node:

- Choose an instance type for the node based on the required performance and capacity

- Select the availability zone where the node will be deployed

- Configure any additional settings, such as custom domain names or VPC settings, as needed

3. Click on **Create peer node** to start the process of creating the new node.

The node creation process can take several minutes. We can monitor its progress in the AMB console.

Removing a node

To remove a node from a member in our network, follow these steps:

1. In the AMB console, navigate to the **Members** tab and select the member whose node we want to remove.

2. Click on the **Peer nodes** tab and select the node to remove.

3. Click on **Remove node** and confirm the action. Note that this action is irreversible and will permanently remove the node and its associated resources from the network.

Monitoring node health

AMB integrates with Amazon CloudWatch, allowing us to monitor the health and performance of our nodes. We can create custom dashboards and set up alarms to receive notifications when specific metrics reach predefined thresholds.

Here are some important metrics to monitor:

- **CPU utilization**: The percentage of the node's CPU capacity being used

- **Memory utilization**: The percentage of the node's memory being used

- **Disk space utilization**: The percentage of the node's disk space being used

- **Network traffic**: The amount of data being sent and received by the node

By following the steps in this section, we can monitor the health and performance of our provisioned nodes using Amazon CloudWatch. Let's create a new dashboard in CloudWatch:

1. Log in to the AWS management console and navigate to the CloudWatch service.

2. In the CloudWatch console, go to the **Dashboards** section and create a new dashboard. Give your dashboard a meaningful name that reflects its purpose:

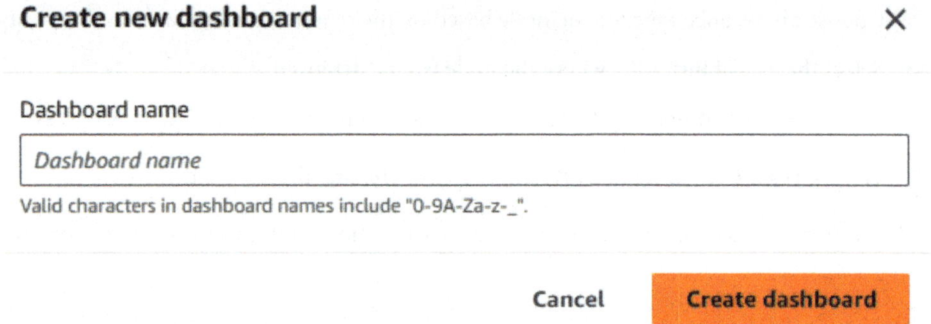

Figure 4.10 – Creating a new CloudWatch dashboard

3. Add widgets to your dashboard to display various AMB metrics. CloudWatch offers different types of widgets, such as graphs, text, or numbers, to represent your data. For each widget, select the AMB metrics you want to monitor. You can display metrics such as block height, transaction rate, CPU utilization, memory usage, and more:

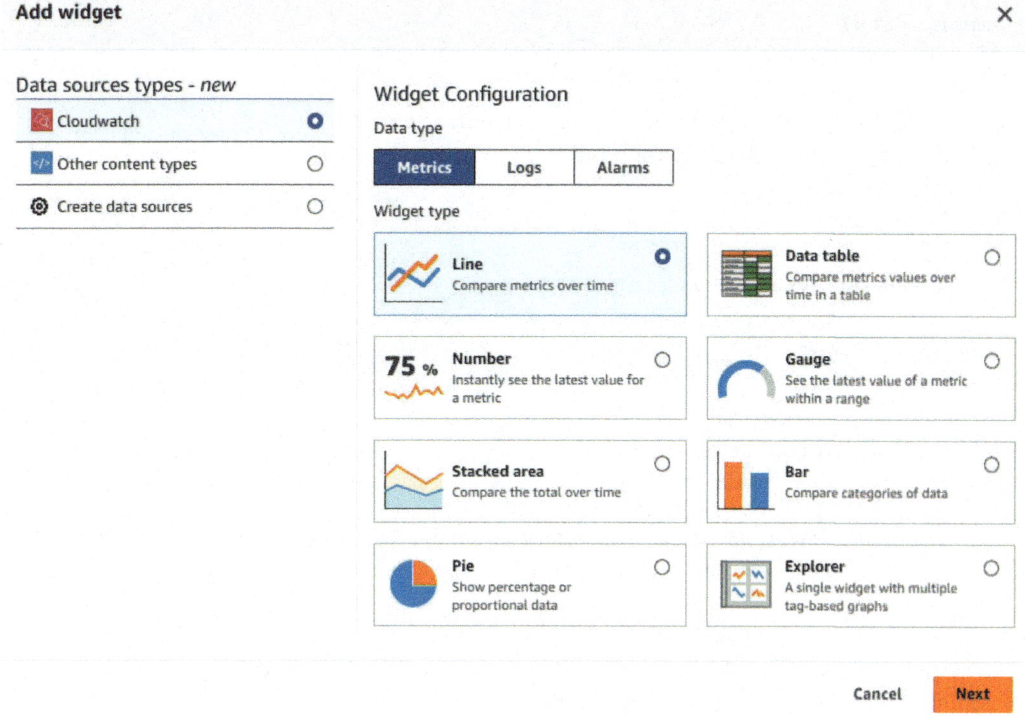

Figure 4.11 – Adding widgets to a CloudWatch dashboard

4. Customize each widget to display the data in the format you prefer (for example, a line chart, stacked area chart, or bar chart). Set the appropriate time range and period for the metrics to ensure you're monitoring the data in real time or over a specific historical period.

Optionally, we can set up CloudWatch Alarms to receive a notification when certain thresholds are breached. For example, we might want to receive an alert if the CPU utilization goes beyond a certain percentage.

Besides the AWS console for creating a new CloudWatch dashboard, we can also utilize a script to automate this process. This can be achieved by using a Python script and Boto3, the AWS SDK for Python. This script assumes you have already configured your AWS credentials and have the necessary permissions to create CloudWatch dashboards and access AMB metrics.

First, we need to install Boto3 if we've not done so already:

```
pip install boto3
```

Then, we must run a Python script to create a CloudWatch dashboard:

```python
import boto3
import json

cloudwatch_client = boto3.client('cloudwatch')
dashboard_name = 'MyAMBDashboard'
dashboard_body = {
    "widgets": [
        {
            "type": "metric",
            "x": 0,
            "y": 0,
            "width": 12,
            "height": 6,
            "properties": {
                "metrics": [
                    ["AWS/ManagedBlockchain", "Invocations",
"NetworkId", "YourNetworkId", "MemberId", "YourMemberId"],
                    [".", "BlockHeight", ".", ".", ".", "."]
                ],
                "view": "timeSeries",
                "stacked": False,
                "region": "us-east-1",
                "stat": "Average",
                "period": 300
            }
        },
        # Add more widgets as needed
    ]
}


# Create or update the dashboard
response = cloudwatch_client.put_dashboard(
    DashboardName=dashboard_name,
    DashboardBody=json.dumps(dashboard_body)
)
```

In this script, we must do the following:

- Replace `YourNetworkId` and `YourMemberId` with our actual AMB network ID and member ID

- Modify the `metrics` array within the widget to include the specific metrics we want to monitor

- Adjust the widget properties, such as `x`, `y`, `width`, `height`, and `region`, as per our requirements

- We can add more widgets by extending the `widgets` array

After creating the dashboard and its widgets, we can set up a CloudWatch alarm using – once again – a Python script with the Boto3 library. Here is an example script that demonstrates how to create an alarm in AWS CloudWatch. This alarm will trigger if a specified metric (for example, `CPUUtilization`) for an AMB resource exceeds a certain threshold:

```python
import boto3

cloudwatch_client = boto3.client('cloudwatch')
alarm_name = 'AMB_CPU_Utilization_Alarm'
namespace = 'AWS/ManagedBlockchain'
metric_name = 'CPUUtilization'
threshold = 70.0  # Set your threshold
evaluation_periods = 1
period = 300  # In seconds
statistic = 'Average'
comparison_operator = 'GreaterThanThreshold'
network_id = 'YourNetworkId'
member_id = 'YourMemberId'
alarm_description = 'Alarm when CPU exceeds 70%'
actions_enabled = False  # Set to True if you want to enable actions
like notifications
alarm_actions = []  # List of action ARNs (e.g., SNS topic ARN) if
actions are enabled

# Creating the alarm
response = cloudwatch_client.put_metric_alarm(
    AlarmName=alarm_name,
    AlarmDescription=alarm_description,
    ActionsEnabled=actions_enabled,
    AlarmActions=alarm_actions,
    MetricName=metric_name,
    Namespace=namespace,
    Statistic=statistic,
    Dimensions=[
        {'Name': 'NetworkId', 'Value': network_id},
```

```
        {'Name': 'MemberId', 'Value': member_id}
    ],
    Period=period,
    EvaluationPeriods=evaluation_periods,
    Threshold=threshold,
    ComparisonOperator=comparison_operator
)
```

In this script, we must do the following:

- Replace `YourNetworkId` and `YourMemberId` with our actual AMB network ID and member ID.

- Set the threshold to the value at which we want the alarm to trigger.

- Use `evaluation_periods` and `period` to define the time window for evaluating the metric.

- Use `actions_enabled` and `alarm_actions` to specify what actions the alarm should trigger. For example, we might want to send a notification via Amazon SNS when the alarm state is reached.

Now that we have CloudWatch integrated into AMB for monitoring the health of the service, we can progress with setting up security and scaling options.

Key considerations for security, scalability, and monitoring

In this section, we will highlight the key considerations for security, scalability, and monitoring when using AMB. We will discuss best practices for securing and scaling the network, as well as how to monitor the network for issues and take appropriate action.

Security considerations

When deploying and managing a blockchain network with AMB, consider the following security best practices:

- **Encryption**: Ensure that data is encrypted both at rest and in transit. AMB automatically encrypts data at rest using AWS KMS and supports encryption in transit using **Transport Layer Security (TLS)**.

- **Access control**: Use AWS IAM to control access to AMB resources. The best practice is to create IAM policies that grant the least privilege necessary to users and roles.

- **Network isolation**: Isolate an AMB network from other networks using Amazon **Virtual Private Cloud** (**VPC**) and VPC endpoints. This will help protect the AMB network from unauthorized access.

- **Auditing and logging**: Enable AWS **CloudTrail** to log API activity for AMB resources. Don't forget to regularly review these logs to detect and respond to potential security incidents.

Scalability considerations

As our blockchain network grows and our application's requirements evolve, we should consider the following best practices for scaling our network:

- **Add or remove nodes** to adjust the number of nodes in our network to meet the demands of our application. Adding nodes can help distribute the workload and improve the network's performance, while removing nodes can help reduce costs when they are no longer needed.

- **Choose the right instance type** to select the appropriate instance type for our nodes based on the performance and capacity requirements of our application. AMB supports various instance types with different levels of compute, memory, and storage resources.

- **Optimize the consensus algorithm** to choose the consensus algorithm that best meets the needs of our application and network size. For example, RAFT is suitable for most use cases, while Kafka is more appropriate for large networks with higher throughput requirements.

Monitoring considerations

Monitoring your AMB network is crucial for detecting and responding to issues promptly. Consider the following best practices for monitoring your network:

- **Use Amazon CloudWatch**: Leverage Amazon CloudWatch to monitor key performance and health metrics for the deployed nodes, such as CPU utilization, memory utilization, and network traffic.

- **Set up alarms**: Configure CloudWatch Alarms to receive notifications when specific metrics reach predefined thresholds. This can help developers proactively identify and address issues before they impact the application.

- **Monitor logs**: Enable AWS CloudTrail to log API activity for AMB resources, and regularly review these logs to detect potential issues or security incidents.

These key considerations on security, scalability, and monitoring provide us with a comprehensive understanding of how to get started with AMB. We are now ready to build, deploy, and manage blockchain applications on AWS. The next section describes a common use case for building a traceability app and its reference architecture, both of which we can deploy on AMB.

Building a tracking application

Now that we have configured an instance of AMB and members, in this section, we'll cover a practical example of how we can build a decentralized application for tracking products on a supply chain. We'll use Hyperledger Fabric since we deployed that in the AMB instance seen in the previous sections of this chapter.

Reference architecture

The supply chain of reference has two members only, for simplicity: a **retailer**, who has products in-store and sells them to the public, and its **supplier**, who manufactures the products and ships them to the retailer. The following diagram shows the AMB components in this solution:

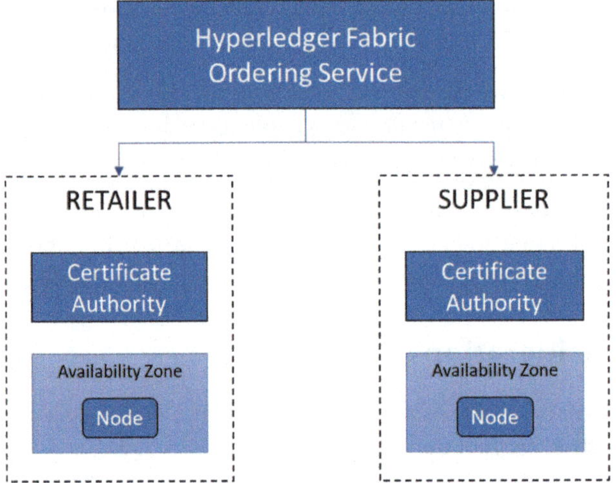

Figure 4.12 – Tracking application solution components in AMB

The ordering service broadcasts new blocks of transactions to the nodes in Hyperledger Fabric. Each participant in the supply chain leverages a certificate authority for managing the authorized access to the network by registered members only. Hyperledger Fabric is a permissioned blockchain where nodes have defined identities that must be registered on the network. Nodes are in separate availability zones for high availability and failover purposes.

Network Setup

By following the steps described in the *Creating a managed blockchain network* section, let's create a new AMB instance called `SupplyChain` and add two members to it: **retailer** and **supplier**. The retailer will be the first member to be added to the network. Once it's been created, the retailer will invite the supplier to join the network by following the steps described in the *Inviting members and managing access* section.

The invite process will make use of an AWS CloudFormation script to facilitate the creation of the necessary resources in AWS. The following YAML file shows the first few lines of a CloudFormation script that deploys the necessary resources for the `SupplyChain` network. The full file is available in the GitHub repository for this chapter:

```
AWSTemplateFormatVersion: 2010-09-09
Description: AMB member invitation acceptance template
Parameters:

   ...

Resources:
  Member:
    Type: AWS::ManagedBlockchain::Member
```

The script consists of two main sections:

- `Parameters`: This defines the AMB network parameters, including network ID, instance type and size, the names of the members, and admin credentials

- `Resources`: This defines the member configuration details (instance type, availability zone, and so on) of each peer node

Save this file as `supplychain-amb.yaml`, and then deploy the described resources by running the `aws cloudformation deploy` command. The `PeerNode1AZ` and `PeerNode2AZ` parameters identify the selected AWS region for deploying the two peer nodes:

```
aws cloudformation deploy --template-file supplychain-amb.yaml
--stack-name amb-supplier --parameter-overrides NetworkId=$NETWORKID
InvitationId=$INVITATIONID MemberName=Supplier AdminUsername=admin
AdminPassword=Admin123 PeerNode1AZ=us-east-1a PeerNode2AZ=us-east-1b
InstanceType=bc.t3.small
```

We can track the progress of the deployment in the CloudFormation service by watching the events' progress in the **Events** tab of the `amb-supplier` stack. Upon deployment, the output should look like this:

```
1    Waiting for changeset to be created..
2    Waiting for stack create/update to complete
3    Successfully created/updated stack - amb-supplier
```

Figure 4.13 – Successfully deploying the AMB network in CloudFormation

The chaincode

Hyperledger Fabric can run chaincode to process business login on-chain. Chaincode is written in JavaScript. For our `SupplyChain` example, we'll write a `product.js` file that contains the following code.

This chaincode verifies that data sent as input to the supply chain is correct before any action is taken. It does so by defining an FSM object (**Finite State Machine**) to ensure that state transitions are valid:

```javascript
const StateMachine = require('javascript-state-machine');
const FSM = new StateMachine.factory({
  init: 'manufactured',
  transitions: [
    { name: 'inspect', from: 'manufactured', to: 'inspected' },
    { name: 'ship', from: 'inspected', to: 'shipped' },
    { name: 'receive', from: 'shipped', to: 'stocked' },
    { name: 'label', from: 'stocked', to: 'labeled' },
    { name: 'sell', from: 'labeled', to: 'sold' },
    { name: 'goto', from: '*', to: function(s) { return s } }
  ]
});
```

Then, the chaincode implements the `ProductsChaincode` class to track the state of products in the supply chain:

```javascript
const ProductsChaincode = class {
  constructor(cid = shim.ClientIdentity) {
    this.clientIdentity = cid;
  }


  assertCanPerformTransition(stub, transition) {
    let requiredAffiliation = 'undefined';
    switch (transition) {
      case 'manufacture':
      case 'inspect':
      case 'ship': requiredAffiliation = 'Supplier'; break;
      case 'receive':
      case 'label':
      case 'sell': requiredAffiliation = 'Retailer';
    }
    this.requireAffiliationAndPermissions(stub, requiredAffiliation,
  transition);
  }
```

This object also ensures that a product can transition from one state to another only when specific conditions are met:

- The `assertCanPerformOperation` method determines which membership affiliations are required for which operations

- It then calls the `requireAffiliationAndPermissions` method to check whether the caller belongs to the specified blockchain member and has the specified permission

The core functionality of transitioning products in the supply chain is executed by the `updateProd-`
`uctState` method, which, as its name implies, updates an existing product as it moves through
the supply chain:

```
async updateProductState(self, stub, args) {
  const productId = args[0];
  const transition = args[1];
  self.assertCanPerformTransition(stub, transition);
  const key = `product_${productId}`;
  const productDataBytes = await stub.getState(key);
  const productData = JSON.parse(productDataBytes.toString());
  const product = new FSM();
  product.goto(productData.state);
  product[transition]();
  productData.state = product.state;
  const now = new Date();
  productData.history = productData.history || {};
  productData.history[product.state] = now.toISOString();
  const stringProductData = JSON.stringify(productData);
  await stub.putState(key, Buffer.from(stringProductData));
  return stringProductData;
}
```

Deployment

Next, we need to build the chaincode into a distribution package. As we need all peer nodes to be able
to access the same package and deploy it, a good and quick approach we can take for distribution is to
copy the package into a common S3 bucket in AWS. The following command packages the chaincode
into the `supplychaincc.tar.gz` archive file, and then uploads it to a shared S3 bucket identified
by the `$BUCKET_NAME` variable:

```
peer lifecycle chaincode package supplychaincc.tar.gz --path $HOME/
chaincode --lang node --label supplychaincc_1.0
sudo chmod 644 supplychaincc.tar.gz
aws s3api put-object --bucket $BUCKET_NAME --key supplychaincc.tar.gz
--body $HOME/supplychaincc.tar.gz --acl bucket-owner-full-control
sudo rm supplychaincc.tar.gz
```

Each member of the AMB network will be able to download the identical `supplychaincc` chaincode
package and install it into their respective peer nodes with the `chaincode install` command:

```
aws s3api get-object --bucket $BUCKET_NAME --key supplychaincc.tar.gz
$HOME/supplychaincc.tar.gz
peer lifecycle chaincode install supplychaincc.tar.gz
CORE_PEER_ADDRESS=$PEER2ENDPOINT peer lifecycle chaincode install
supplychaincc.tar.gz
```

The next deployment step is approval. Hyperledger Fabric requires that all nodes in a network approve the deployment of chaincode. This can be accomplished with the `chaincode approve-formyorg` command:

```
peer lifecycle chaincode approveformyorg -o $ORDERER --channelID
mainchannel --name supplychaincc --version 1.0 --sequence 1 --init-
required --package-id $SUPPLYCHAIN_CC_PACKAGE_ID --tls --cafile $HOME/
managedblockchain-tls-chain.pem
```

We can also use the `chaincode checkcommitreadiness` command to verify that the chaincode has been approved by all members and is ready for its final commit to the network:

```
peer lifecycle chaincode checkcommitreadiness -o $ORDERER --channelID
mainchannel --name supplychaincc --version 1.0 --init-required
--sequence 1 --tls --cafile $HOME/managedblockchain-tls-chain.pem
--output json
```

Lastly, to make the chaincode executable on the network for each member, one member should commit it using the `chaincode commit` command:

```
peer lifecycle chaincode commit -o $ORDERER --channelID mainchannel
--name supplychaincc --version 1.0 --sequence 1 --init-required --tls
--cafile $HOME/managedblockchain-tls-chain.pem
```

Our chaincode requires initialization. This can be achieved by running the `chaincode invoke` command with the `--isInit` flag for the first time:

```
peer chaincode invoke -C mainchannel -n supplychaincc --isInit -c
'{"Args": ["init"]}' -o $ORDERER --cafile $HOME/managedblockchain-tls-
chain.pem --tls --waitForEvent
```

At this point, the chaincode has been published on the AMB network and can be operated. Likely, the supply chain application will have a frontend for user interaction, which will consume the chaincode for managing the state transition of products by all relevant and authorized parties.

Summary

This chapter introduced AMB, a service that simplifies the process of building, deploying, and managing blockchain networks on AWS. We covered how to create a network, configure its settings, and invite members. We also discussed how to deploy and manage nodes, as well as how to secure, scale, and monitor the network. Finally, we provided some best practices for developing blockchain applications on AWS using AMB.

In the next chapter, we will explore how to host a blockchain network on Elastic Kubernetes Service in AWS.

Further reading

- AMB documentation:

  ```
  https://docs.aws.amazon.com/managed-blockchain/
  ```

- *Get Started Creating a Hyperledger Fabric Blockchain Network Using Amazon Managed Blockchain (AMB)*:

  ```
  https://docs.aws.amazon.com/managed-blockchain/latest/hyperledger-
  fabric-dev/managed-blockchain-get-started-tutorial.html
  ```

- Boto3 documentation:

  ```
  https://boto3.amazonaws.com/v1/documentation/api/latest/index.
  html
  ```

5

Hosting a Blockchain Network on Elastic Kubernetes Service

In this chapter, we will delve into the fascinating world of blockchain technology and cloud-based services, focusing on the powerful combination of **Elastic Kubernetes Service (EKS)** and **Hyperledger Fabric**.

We will cover the following topics:

- Introduction to Hyperledger Fabric on EKS

- Creating an EKS cluster for hosting the Hyperledger Fabric blockchain

- Deploying a Hyperledger Fabric blockchain network on EKS

- Key considerations – Security, scaling, and monitoring

- Testing and troubleshooting a Hyperledger Fabric blockchain network on EKS

Technical requirements

Hosting a blockchain network on Amazon EKS involves a set of technical requirements and considerations to ensure optimal performance, security, and scalability. The following requirements form a comprehensive checklist for hosting a blockchain network on Amazon EKS:

- **Understanding Amazon EKS and Kubernetes**: Familiarity with the architecture of Amazon EKS, including how it manages Kubernetes clusters and integrates with other **Amazon Web Services (AWS)**, as well as a fundamental understanding of Kubernetes concepts, including pods, services, deployments, and stateful sets, is required

- **AWS account and IAM configuration**: An active AWS account is necessary to access Amazon EKS and other related services and properly configured IAM roles for EKS, ensuring permissions for creating and managing Kubernetes clusters and interacting with other AWS resources

This chapter will expand on key concepts of network setup and configuration and the creation of an EKS cluster. It's important to have basic knowledge of how to operate in AWS for the configuration of a **Virtual Private Cloud** (**VPC**) for hosting the EKS cluster, ensuring network isolation and security. This will be achieved by introducing adequate subnetting within the VPC to separate resources and manage traffic effectively, with the use of an **Internet Gateway** for external connectivity, especially for public blockchain networks.

Introduction to Hyperledger Fabric on EKS

Amazon's EKS is a service offered by AWS that allows users to manage Kubernetes. This service facilitates the deployment, management, and scaling of applications that are containerized using Kubernetes. EKS provides a solution that is secure, scalable, and highly available for hosting blockchain networks, including but not limited to Hyperledger Fabric.

Hyperledger Fabric is a permission-based, modular, and extensible blockchain platform designed for enterprise use. It supports smart contracts written in various languages and offers a customizable consensus mechanism. This chapter focuses on hosting a Hyperledger Fabric network on EKS, but the approach can be adapted for other blockchain platforms.

Architecture and components

At a high level, the software architecture of Hyperledger Fabric includes the following components:

Figure 5.1 – High-level components of Hyperledger Fabric

The key components together provide a solid platform for running solutions in a permission-based blockchain network:

- **Peer nodes**: These are nodes that commit transactions and maintain the state and a copy of the ledger. Peers can be endorsers, committing peers, or both.

- **Ordering service**: The component responsible for achieving consensus on the order of transactions and packaging them into blocks.

- **Chaincode**: Business logic is implemented in chaincode (the Hyperledger term for smart contracts) and is run by peers.

- **Membership Service Provider (MSP)**: This component handles identity and membership services, ensuring authenticated and authorized access to the network.

- **Channels**: Data partitioning mechanism that allows for a group of participants to create a separate ledger of transactions.

Hyperledger Fabric supports private transactions and confidential contracts, unlike many blockchain solutions. Its modular architecture allows network designers to plug in their preferred implementations for components, which are optimized for a high transaction rate and low latency. The platform can scale to a large number of nodes and transactions in a network, enriched by a query capability against the state database.

In terms of integration with Kubernetes, Hyperledger Fabric enjoys the following features, which will be expanded on in the next sections:

- **Dynamic scalability**: Kubernetes facilitates the dynamic scaling of the Fabric network

- **Container orchestration**: The efficient management of Fabric components (peers, orders, etc.) as containerized applications.

- **Resilience and high availability**: Kubernetes ensures high availability of the Fabric network with features such as self-healing and automated rollouts/rollbacks.

EKS provides a robust, scalable, and secure environment for containerized applications, and Hyperledger Fabric offers a flexible and secure platform for building enterprise-grade blockchain applications. Their integration can yield powerful solutions, combining the efficiency and scalability of Kubernetes with the security and flexibility of Hyperledger Fabric.

The next sections will expand on the strong characteristics of the integration of Hyperledger Fabric with EKS.

Dynamic scalability

Dynamic scalability refers to the ability of the system to automatically adjust its resources to meet changing demands. In the context of Hyperledger Fabric on EKS, this means the capability to scale the blockchain network components up or down based on the workload.

The following figure shows a four-step flow of operations that will enhance the scalability of a Hyperledger Fabric setup in EKS:

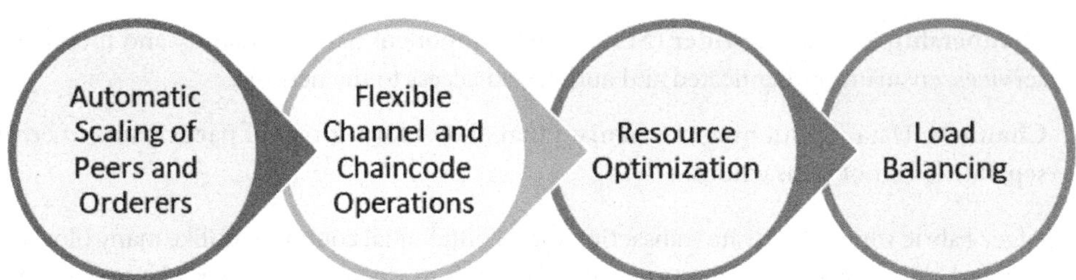

Figure 5.2 – Steps to enhance dynamic scalability of Hyperledger Fabric in EKS

These steps can be summarized as follows:

1. Kubernetes can automatically increase or decrease the number of peer and order-placing nodes in a Hyperledger Fabric network. This is typically managed through Kubernetes **Horizontal Pod Autoscaler (HPA)**, which adjusts the number of pod replicas based on CPU utilization or other select metrics.

2. As the demand for different channels or chaincodes varies, Kubernetes can **dynamically allocate resources** to maintain optimal performance.

3. EKS ensures the **optimal usage of underlying AWS resources**, helping to manage costs and performance effectively.

4. EKS can **distribute network traffic** across various nodes evenly, ensuring that no single node becomes a bottleneck.

To give a tangible example, think about how a financial blockchain application for cross-border payments may have transaction volumes that can vary dramatically. During high transaction periods, such as fiscal year-ends or during significant global events, the network can automatically scale to handle the increased load, ensuring smooth and uninterrupted transaction processing.

In supply chain management, there might be times, such as holiday seasons or product launches, when tracking activities surge significantly in the supply chain network. EKS can dynamically scale the Hyperledger Fabric network to handle the increased tracking and verification requests, maintaining the performance and responsiveness of the system.

One more example in healthcare data management is where the demand for data access and recording can fluctuate, especially during public health crises. A Hyperledger Fabric network on EKS can adjust its capacity to handle these fluctuations, ensuring that patient data is consistently and quickly accessible to authorized individuals.

In each of these examples, the key benefit of using Hyperledger Fabric on Amazon EKS is its ability to respond to varying loads with minimal manual intervention, ensuring that the blockchain network remains efficient, cost-effective, and highly available.

Later in this chapter, we'll see an example of a script template that can be used to introduce dynamic scalability in our deployed Hyperledger Fabric network on EKS.

Container orchestration

The orchestration of the containers of Hyperledger Fabric using EKS involves managing the deployment, scaling, and operation of Hyperledger Fabric components within the Kubernetes ecosystem. The following figure summarizes the five key components to consider for Hyperledger Fabric container orchestration using EKS:

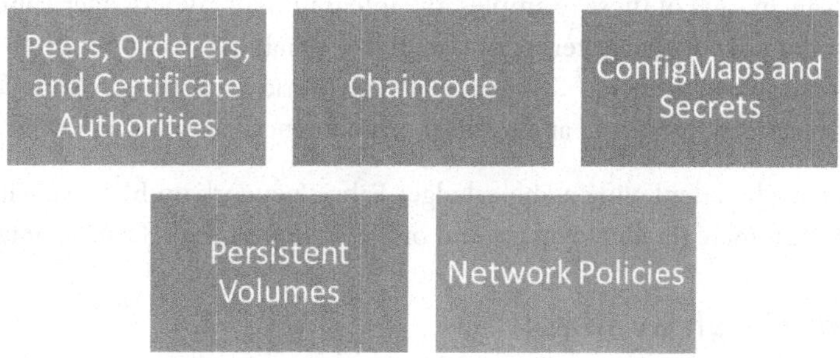

Figure 5.3 – Key components of container orchestration for Hyperledger Fabric with EKS

Let's elaborate on these key components:

- **Peers, order placers, and Certificate Authorities (CAs)**: These are the core components that are containerized and managed as Kubernetes pods.

- **Chaincode**: This is the equivalent of smart contracts in Hyperledger Fabric; it is code deployed and executed within Docker containers and managed by peer nodes.

- **ConfigMaps and secrets**: These are used for storing and managing configuration data and sensitive information such as credentials.

- **Persistent volumes**: This ensures data persistence for blockchain ledgers and state databases (CouchDB or LevelDB).

- **Network policies**: These define how pods communicate with each other, enhancing network security.

Returning to the example in a financial transaction network, imagine a scenario where a multinational bank deploys a Hyperledger Fabric network on EKS for processing international transactions. A potential implementation may consider that peer and order-placing nodes be entirely containerized and managed through EKS, ensuring high availability and scalability. During peak transaction periods, EKS automatically scales up the resources to handle increased loads, maintaining transaction processing efficiency.

In a supply chain tracking platform, a global supply chain company uses Hyperledger Fabric on EKS to track the movement of goods in real time. Each step in the supply chain is recorded on the blockchain. The implementation of such a solution may look at Kubernetes for orchestrating the entire network and managing numerous peer nodes deployed worldwide. EKS ensures data consistency and availability, even with the dynamic nature of supply chain logistics.

A third example in healthcare data management, specifically a healthcare consortium that implements a Hyperledger Fabric network on EKS for secure and confidential patient data sharing, would implement EKS to manage the deployment of Fabric's components, handling sensitive patient data with strict privacy controls. The system would scale as per the incoming data traffic, ensuring the efficient management of large volumes of healthcare data.

As we can appreciate, in each of these examples, the integration of Hyperledger Fabric with Amazon EKS offers significant advantages in terms of scalability, reliability, and operational efficiency. The container orchestration capabilities of EKS ensure that the Fabric network can dynamically adapt to the specific demands of various real-world applications, from financial services to government operations.

Later in this chapter, when deploying a Hyperledger Fabric network on EKS, we will look at specific script templates to automate the deployment and orchestration of the relevant containers.

Resilience and high availability

Resilience and high availability are critical aspects of deploying Hyperledger Fabric on EKS. This approach ensures that the blockchain network remains operational and robust against various types of failures, including hardware malfunctions, network issues, and unexpected surges in demand. The following figure depicts critical capabilities for enhancing the resilience and high availability of Hyperledger Fabric on EKS:

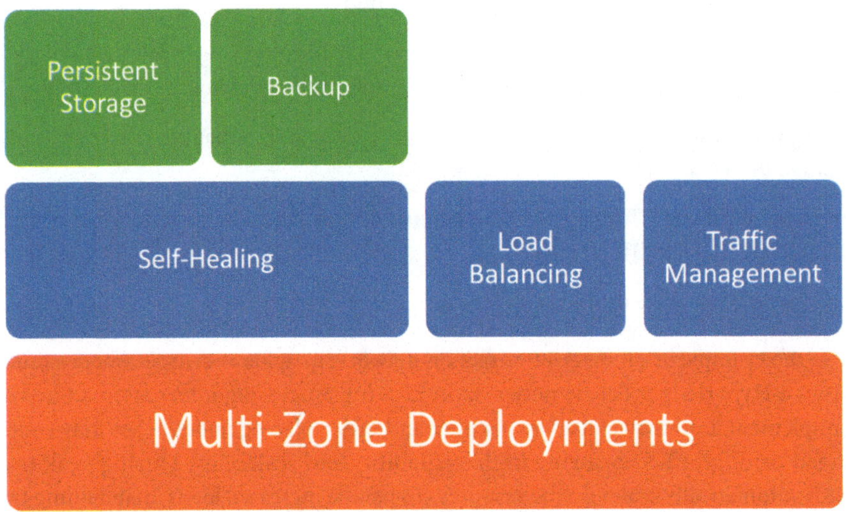

Figure 5.4 – Capabilities for enhancing the resilience and high availability of Hyperledger Fabric

For a start, EKS supports deployment across **Availability Zones** (**AZ**). Deploying the EKS cluster across multiple AZs in AWS ensures that the failure of one AZ doesn't impact the blockchain network. This multi-AZ approach provides redundancy and high availability.

EKS can also automatically **detect and replace unhealthy nodes** (peer nodes, order-placing nodes, etc.) in the Hyperledger Fabric network. This Kubernetes' self-healing mechanisms ensure minimal downtime.

A critical capability is **Elastic Load Balancing** (**ELB**): ELB can distribute incoming network traffic across multiple instances of Hyperledger Fabric's components, preventing any single instance from being overwhelmed.

One last comment about storage comes in two flavors: first, reliable data storage can be delivered by integrating AWS services such as **Elastic Block Store** (**EBS**) and **Elastic File System** (**EFS**). Even if a peer node fails, the ledger data remains intact. Secondly, we may want to implement automated backups, that is, regular backups of the blockchain state using **AWS backup**, for enhanced data resilience.

Specific script examples are provided later in this chapter to configure high availability for a Hyperledger Fabric implementation in EKS.

Going back to our real-world examples, how can we improve the high availability of a global trade finance platform that uses Hyperledger Fabric on EKS for processing international trade transactions? In this scenario, the EKS cluster would be spread across multiple AZs, ensuring that the trade finance transactions continue smoothly even if one AZ faces an outage. Regular data backups and automatic failover mechanisms ensure data integrity and continuous operation.

As seen before, the retail supply chain solution uses Hyperledger Fabric on EKS to manage its supply chain and inventory. The network's high availability would ensure the uninterrupted tracking of goods across different geographical locations. In case of a node failure, Kubernetes' self-healing capabilities quickly restore the service with minimal disruption.

The last example is for a healthcare data exchange made of a consortium of hospitals that implement a blockchain network on EKS for secure and confidential patient data exchange. In this case, the implementation of EKS would automatically handle node failures, ensuring that patient data is always accessible. ELB efficiently manages traffic during peak times, such as during health crises, ensuring system responsiveness.

In each of these examples, the combination of Hyperledger Fabric with Amazon EKS offers a robust and resilient infrastructure crucial for maintaining continuous operations in critical applications such as finance, healthcare, and supply chain management. The ability of EKS to provide high availability, along with its self-healing and load-balancing capabilities, makes it an ideal choice for hosting enterprise-grade blockchain networks.

Now that we have understood the value that EKS adds to a blockchain solution, and specifically to an implementation of Hyperledger Fabric, let's progress to the next section to understand the creation of an EKS cluster for hosting a Hyperledger Fabric blockchain.

Creating an EKS cluster for hosting the Hyperledger Fabric blockchain

Creating an Amazon EKS cluster to host a Hyperledger Fabric network involves several steps, from setting up the AWS **Command-Line Interface (CLI)** and EKS CLI to configuring the Kubernetes cluster itself. Step-by-step, let's complete the following tasks using an active AWS account:

1. Install and configure the **AWS CLI** with a user that has the necessary permissions, which are `AmazonEKSAdminPolicy`, `AmazonEC2FullAccess`, `IAMFullAccess`, `AmazonS3FullAccess`, and `AmazonVPCFullAccess`. The latest version of the AWS CLI for your platform (Windows, macOS, or Linux) can be found here: `https://docs.aws.amazon.com/cli/latest/userguide/getting-started-install.html`.

2. Install **eksctl**, a simple CLI tool for creating clusters on EKS. It simplifies much of the cluster creation process. Instructions on how to install eksctl can be found here: `https://github.com/eksctl-io/eksctl`.

3. Create an IAM role that EKS can assume to create AWS resources for Kubernetes clusters. This is typically handled by eksctl automatically but can also be done manually through the **AWS Management Console** by opening the IAM console at `https://console.aws.amazon.com/iam`.

4. Use eksctl to create an EKS cluster. This command creates a cluster with all the default settings, including default node types and counts. Don't forget to specify the region of preference after the `--region` parameter:

    ```
    eksctl create cluster --name my-hlf-cluster --version 1.21
    --region us-west-2 --nodegroup-name my-nodes --node-type
    t3.medium --nodes 3 --nodes-min 1 --nodes-max 4 --managed
    ```

    More specifically, the parameters for the `create cluster` command are the following:

    - `--name`: The name of the EKS cluster

    - `--version`: Kubernetes version

    - `--region`: The AWS region for the EKS cluster

    - `--nodegroup-name`: The name of the node group

    - `--node-type`: The type of EC2 instances for the nodes being deployed in the cluster

    - `--nodes`: The initial number of nodes

    - `--nodes-min`: The minimum number of nodes to allocate (for auto-scaling)

    - `--nodes-max`: The maximum number of nodes allowed (for auto-scaling)

    - `--managed`: Specifies that we are creating a managed node group.

5. After the cluster is created, we need to configure **kubectl** and enable cluster communication. This command updates the `kubeconfig` on the virtual machine to use the new EKS cluster:

```
aws eks --region us-west-2 update-kubeconfig --name my-hlf-
cluster
```

At this point, the EKS cluster is ready; it may take a few minutes for the resources to actually be ready. Communication is enabled. In the next section, we'll focus on the steps to deploy Hyperledger Fabric in the EKS cluster that we have just created.

Deploying a Hyperledger Fabric blockchain network on EKS

Now that the EKS cluster is set up, we can start deploying the Hyperledger Fabric components. This involves preparing the relevant Docker images for Fabric components for the peers, order-placing, and CA nodes. The official Hyperledger Fabric images can be obtained at `https://hyperledger-fabric.readthedocs.io/en/release-2.2/install.html`.

In addition to creating the Kubernetes deployments and services for each component, we will also need to set up persistent storage for data persistence, ideally using AWS EBS or EFS.

Persistent Storage

Setting up persistent storage for a Hyperledger Fabric network in Amazon EKS using EBS involves creating **Persistent Volumes (PVs)** and **Persistent Volume Claims (PVCs)** in a Kubernetes cluster. This setup ensures data persistence for key components such as peers, order-placing, and CAs nodes across any restarts and failures.

First, we need to manually create an EBS volume in the same AWS region as the EKS cluster or use dynamic provisioning with a StorageClass. For dynamic provisioning (the recommended approach), we can define a StorageClass specifically for EBS and then create PVCs that automatically provision EBS volumes.

To define a StorageClass for dynamic storage provisioning, create a file named `ebs-storageclass.yaml` with the following content:

```
kind: StorageClass
apiVersion: storage.k8s.io/v1
metadata:
  name: ebs-gp2
provisioner: kubernetes.io/aws-ebs
parameters:
  type: gp2
```

```
    fsType: ext4
  reclaimPolicy: Retain
  allowVolumeExpansion: true
```

This StorageClass specifies the use of the `gp2` volume type (general purpose SSD) and sets the `reclaimPolicy` to `Retain`, meaning the volume will be retained after the PVC is deleted.

We'll then use kubectl to apply this StorageClass:

kubectl apply -f ebs-storageclass.yaml

For each Hyperledger Fabric component requiring persistent storage (e.g., peers and order placers), we now need to create a PVC. Here's an example PVC for a Fabric peer:

```
kind: PersistentVolumeClaim
apiVersion: v1
metadata:
  name: peer0-pvc
spec:
  accessModes:
    - ReadWriteOnce
  storageClassName: ebs-gp2
  resources:
    requests:
      storage: 10Gi
```

This PVC requests a 10 GiB volume using the `ebs-gp2` StorageClass created earlier.

Additionally, in this case, we will use **kubectl** to apply this PVC:

kubectl apply -f peer0-pvc.yaml

When deploying Hyperledger Fabric components as pods, we can reference the PVC for their storage requirements. The following YAML snippet is an example of a peer deployment. This configuration mounts the EBS-backed persistent volume at `/var/hyperledger/production` in the peer container, ensuring that ledger data are stored in the persistent volume:

```
volumes:
  - name: peer-storage
    persistentVolumeClaim:
      claimName: peer0-pvc
containers:
  - name: peer
    image: hyperledger/fabric-peer:latest
```

```
    volumeMounts:
    - mountPath: /var/hyperledger/production
      name: peer-storage
```

We can now progress to the next section and set up the Hyperledger Fabric components.

Fabric components

With the StorageClass and PVCs defined and the deployment configurations updated to use these PVCs, we can now deploy the Hyperledger Fabric components on EKS. The EBS volumes will be automatically provisioned and attached to the pods requiring persistent storage.

Setting up and orchestrating a Hyperledger Fabric network on EKS involves several steps, including creating Kubernetes manifests for each component of the Fabric network (such as peers, order placers, and certificate authorities). The following is a basic example to illustrate how we might set up a simple peer node in a Hyperledger Fabric network using a Kubernetes deployment and service manifest.

Let's start by downloading the deployment and service YAML configurations in the `hlf-peer-deployment.yaml` and `hlf-peer-service.yaml` files from the GitHub repository for this chapter. The files are too long to reproduce in line in this chapter.

We will use kubectl to deploy the peer node with the following bash commands:

```
kubectl apply -f hlf-peer-deployment.yaml
kubectl apply -f hlf-peer-service.yaml
```

The deployment file makes use of the latest available version of a Docker image `hyperledger/fabric-peer:latest`. A few environment variables, such as `CORE_PEER_ID` and `CORE_PEER_ADDRESS`, also need to be set according to the specific network configuration.

Creating an EKS cluster for Hyperledger Fabric is just the beginning. The real challenge (and fun!) comes with managing, scaling, and ensuring the security of the network as you start deploying chaincode and handling transactions.

Remember, orchestrating a Hyperledger Fabric network on EKS can get complex, especially as we scale up the network, add more organizations, or implement advanced features such as channel participation and chaincode deployment. It is essential to thoroughly understand both Kubernetes and Hyperledger Fabric to manage such a deployment effectively.

In the next section, we will have a look at a few of these key management considerations, specifically about security, high availability, and scalability.

Common deployment challenges

Deploying a Hyperledger Fabric blockchain network on EKS offers a robust platform for blockchain applications. However, the process can present challenges, especially concerning network setup, configuration, and management. The following table presents some common deployment challenges and their solutions:

Challenge	Problem	Solution
Managing cryptographic materials	Hyperledger Fabric requires a meticulous setup of cryptographic materials for peers, order placers, and CAs. The mismanagement of these materials can lead to security vulnerabilities or network failure.	Utilize Hyperledger Fabric CA to streamline the generation, renewal, and revocation of certificates. Store cryptographic materials securely using Kubernetes Secrets and automate their deployment and rotation to minimize human error. Implement strict access controls and auditing to track the use of these materials.
Network configuration and channel setup	Configuring the network and setting up channels can be complex, requiring precise co-ordination between different network members and careful management of configuration files.	Use comprehensive scripts or automation tools such as Ansible, Helm charts, or Kubernetes operators, which are specifically designed for Hyperledger Fabric to automate the deployment and configuration of network components. Ensure that configuration files are version-controlled and reviewed for accuracy before deployment.
Persistent storage for ledger data	Blockchain applications require persistent storage to maintain the ledger and state data. Kubernetes pods are ephemeral, which could lead to data loss if not properly managed.	Implement PVCs in Kubernetes for ledger storage, ensuring data persists across pod restarts and updates. Choose a reliable storage class that matches your performance and redundancy requirements, such as Amazon EBS volumes for stateful data.
Network scalability and performance tuning	As the network grows, it may face scalability issues, with increased transaction volumes leading to performance bottlenecks.	Monitor network performance regularly using tools such as Prometheus and Grafana. Scale resources vertically (upgrading existing resources) or horizontally (adding more peers/order placers) based on the observed load. Use Kubernetes HPA to scale your components automatically. Optimize chaincode for efficiency and evaluate channel architecture to distribute load effectively.

Ensuring high availability and disaster recovery	Downtime in a block-chain network can disrupt operations, while data loss can compromise the network's integrity.	Deploy multi-zone Kubernetes clusters to ensure high availability of the Hyperledger Fabric network. Use StatefulSets for stateful applications such as Fabric peers and order placers to manage pod identity and storage across restarts. Implement backup and recovery procedures for blockchain data and configurations, leveraging cloud storage solutions for backups.
Network security and access control	Ensuring the security of the blockchain network and controlling access to sensitive operations and data is critical.	Use Kubernetes network policies to control traffic flow between pods and enforce least privilege access. Integrate Hyperledger Fabric's built-in access control features with Kubernetes **Role-Based Access Control (RBAC)** to manage permissions for different network participants. Encrypt data in transit using TLS certificates and data at rest using Kubernetes Secrets.

Table 5.1 – Challenges to and solutions for the deployment of Hyperledger Fabric in EKS

Deploying and managing a Hyperledger Fabric network on EKS requires careful planning and attention to detail. By addressing common challenges with thoughtful solutions, organizations can build secure, scalable, and resilient blockchain solutions that leverage the best of Kubernetes and Hyperledger Fabric.

Key considerations – Security, scaling, and monitoring

When hosting a Hyperledger Fabric network on EKS, consider the following best practices for improving the security posture and business continuity of the solution. Specifically, let's look at best practices for security, high availability, and scalability.

Enterprise deployments

Before we dive into each aspect from a configuration perspective, let's analyze a few real-world examples that demonstrate how these considerations are addressed in enterprise deployments.

Security in Hyperledger Fabric – Trade finance platform by we.trade

we.trade is a blockchain-based trade finance platform developed by a consortium of banks to simplify and secure international trade transactions for SMEs. The platform leverages Hyperledger Fabric to manage, track, and protect trade transactions between buyers, sellers, and banks.

The key security measures put in place in the Hyperledger Fabric deployment are the following:

- **Private channels**: we.trade uses Hyperledger Fabric's private channels to ensure transaction privacy between trading parties, enabling the confidential and secure exchanges of information.

- **MSP**: The platform utilizes Fabric's MSP for robust identity management, ensuring that only authorized users can access the network and perform transactions, significantly reducing the risk of fraud and unauthorized access.

- **Smart contract security**: we.trade implements comprehensive smart contract (chaincode) security practices, including thorough testing and code reviews to prevent vulnerabilities and ensure that business logic is accurately and securely executed.

The we.trade case study by IBM is available at this URL: `https://www.ibm.com/case-studies/wetrade-blockchain-fintech-trade-finance`.

High availability in Hyperledger Fabric – The Food Trust Network by IBM

IBM's **Food Trust Network** is a blockchain solution that provides traceability in the food supply chain, enhancing food safety and reducing waste. It's built on Hyperledger Fabric and involves various stakeholders, including producers, wholesalers, retailers, and regulators.

High-availability measures have been put in place for the following critical areas:

- **Multi-region Kubernetes clusters**: To ensure high availability, the Food Trust Network is deployed across multiple geographic regions using Kubernetes clusters. This setup protects against regional outages, ensuring the network remains operational.

- **Replicated ordering service**: The network uses a replicated ordering service, ensuring that even if one instance fails, transaction ordering can continue without interruption, which is critical for maintaining continuous network operations.

- **Disaster recovery plan**: IBM's Food Trust Network has a comprehensive disaster recovery plan, including regular backups and a strategy for rapid restoration in case of significant failures, ensuring data integrity and minimal downtime.

The Food Trust Network case study by IBM is available at this URL: `https://www.ibm.com/products/supply-chain-intelligence-suite/food-trust`

Scalability in Hyperledger Fabric – TradeLens by Maersk and IBM

TradeLens was a global trade digitization platform powered by blockchain technology, developed jointly by Maersk and IBM. The platform facilitated secure and transparent container logistics, making it possible for all trading partners to collaborate efficiently.

A few scalability measures were applied to reach a global market:

- **Horizontal scaling**: TradeLens was designed to scale horizontally by adding more peers to the network to handle the increased load, leveraging Kubernetes for dynamic scaling based on demand.

- **Partitioning and channel strategy**: To manage scalability, TradeLens employed a smart partitioning strategy, using channels to segregate transactional data among different participants and ensure that the network can efficiently handle a large volume of transactions without performance degradation.

- **Performance optimization**: The platform focuses on chaincode performance optimization, ensuring that smart contracts execute efficiently to handle high transaction throughput, which is vital for the scalability of the network.

While TradeLens successfully developed a viable platform based on Hyperledger Fabric, unfortunately, the need for full global industry collaboration was not achieved. As a result, TradeLens did not reach the level of commercial viability necessary to continue operations and meet financial expectations. The business was shut down in 2023. More information can be found in this article: `https://www.maersk.com/news/articles/2022/11/29/maersk-and-ibm-to-discontinue-tradelens`.

These real-world examples demonstrate how enterprises deploying Hyperledger Fabric can address the critical considerations of security, high availability, and scalability. By leveraging Hyperledger Fabric's features, such as private channels for security, support for multi-region deployments for high availability, and strategies for horizontal scaling and performance optimization, enterprises can build blockchain solutions that meet the demanding requirements of large-scale, mission-critical applications.

Security

From an infrastructure perspective, a strong good practice is to use **network policies** to isolate and secure communication between components.

At the application level, we may want to protect sensitive data, such as certificates and keys with **Kubernetes secrets**, and implement **RBAC** for managing permissions.

However, this is just scraping the surface of a robust security practice. Securing a Hyperledger Fabric deployment on EKS involves multiple layers of security, including network configurations, access controls, data encryption, and operational security practices. The **security onion** in the following figure describes the multiple layers of security to implement:

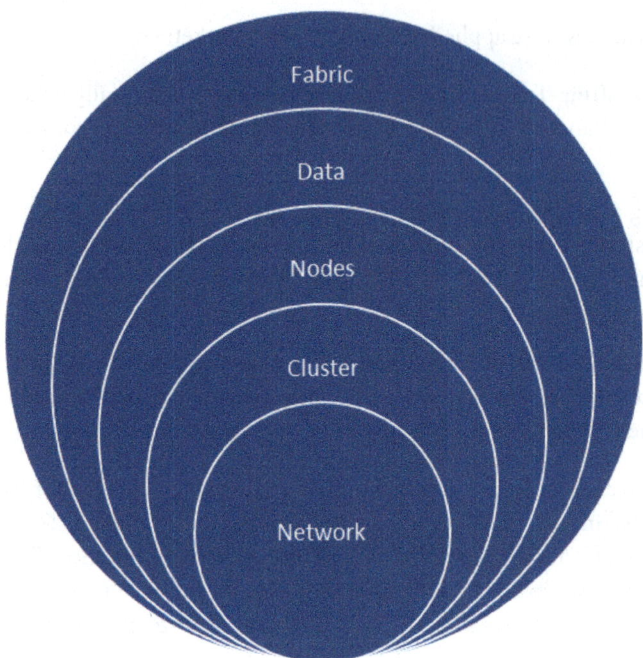

Figure 5.5 – The security onion

Here's a comprehensive approach to hardening an Hyperledger Fabric network:

- **Network security**: Utilize the Amazon VPC **Container Network Interface** (**CNI**) plugin for Kubernetes to assign VPC IP addresses to Kubernetes pods, enabling the use of VPC security groups and network ACLs to control pod traffic.

  Define Kubernetes network policies to control the flow of traffic between pods within the EKS cluster, ensuring that only authorized services can communicate with the Hyperledger Fabric components.

  Configure the EKS cluster for private access, ensuring that the Kubernetes API endpoint is not accessible from the internet. Use VPC endpoints and VPN connections for administrative access.

- **Cluster-level security**: Assign fine-grained roles to the Kubernetes pods using **IAM Roles for Service Accounts** (**IRSAs**) to limit the AWS permissions that applications can use, reducing the risk of unauthorized access to AWS resources.

  Enable logging for the EKS cluster control plane, including audit and API logs, to monitor and record all activities. Use Amazon CloudWatch Logs for storage and analysis.

- **Node security**: Use the latest **Amazon Machine Images** (**AMIs**) for worker nodes to ensure they include the latest security patches, and regularly update the EKS clusters and worker nodes to the latest versions to benefit from security fixes.

- **Data security**: Encrypt all data at rest by using Kubernetes secrets to manage sensitive information such as credentials and config files. For enhanced security, encrypt secrets using the AWS **Key Management Service (KMS)**.

  Ensure that the EBS volumes used by Hyperledger Fabric components are encrypted using AWS KMS.

  Encrypt data in transit by enabling **Transport Layer Security (TLS)** for all communication within the Hyperledger Fabric network. This includes peer-to-peer communications, client applications communicating with the network, and communication between services such as CA, order placers, and peers.

  Use **AWS Certificate Manager (ACM)** or your own public key infrastructure (PKI) to manage certificates for TLS.

- **Hyperledger Fabric security**: Utilize Hyperledger Fabric's Certificate Authority for issuing and managing identities within your Fabric network. Ensure that identities are issued based on strong authentication practices.

Security best practices include regularly auditing the production environment using tools such as AWS Security Hub, Amazon Inspector, and third-party Kubernetes security solutions to ensure compliance with security best practices and standards. Implement a comprehensive backup and disaster recovery plan for your Hyperledger Fabric data stored on EBS volumes and other persistent storage options.

High availability

Configuring resilience and high availability for a Hyperledger Fabric network in EKS involves several Kubernetes concepts such as multi-AZ deployments, pod disruption budgets, liveness and readiness probes, and persistent storage. The following figure is a conceptual illustration that demonstrates how these can be configured:

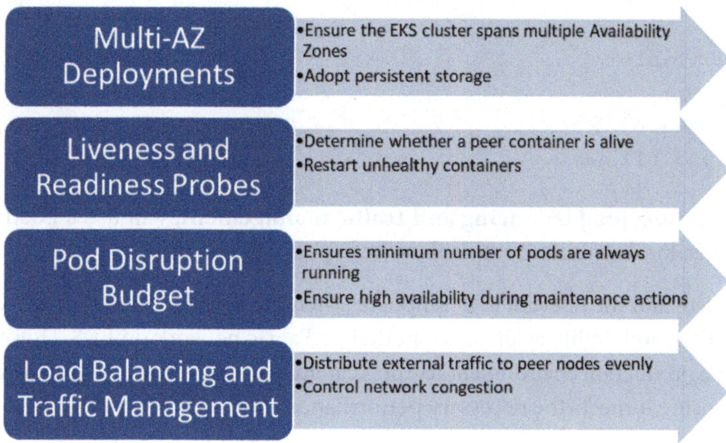

Figure 5.6 – Techniques for ensuring high availability and resilience

With **multi-AZ deployments**, we can ensure that the EKS cluster spans multiple availability zones for higher availability that is not dependent on a single deployment zone. This is part of the EKS cluster setup and can be specified during the cluster creation phase in AWS.

Persistent storage using StatefulSets and PVs ensures that each peer node has its own persistent storage, which is crucial for maintaining the state and data of the blockchain.

Liveness and readiness probes help Kubernetes determine whether a peer container is alive and ready to accept traffic, ensuring that unhealthy containers are restarted and don't impact network performance. A liveness probe can be configured with a YAML script as follows. The probe, after a 15-second initial delay (`initialDelaySeconds` variable), will ping (`peer node status` command) the peer node every 20 seconds (`periodSeconds` variable):

```
livenessProbe:
  exec:
    command:
    - sh
    - -c
    - peer node status
  initialDelaySeconds: 15
  periodSeconds: 20
```

A **Pod Disruption Budget** (**PDB**) ensures that a certain number of pods are always running, even during maintenance and updates. In this case, a simple YAML configuration will suffice for defining the minimum number of pods (`minAvailable` variable):

```
apiVersion: policy/v1beta1
kind: PodDisruptionBudget
metadata:
  name: hlf-peer-pdb
spec:
  minAvailable: 1
  selector:
    matchLabels:
      app: hlf-peer
```

The last characteristic, **load balancing and traffic management**, will use a LoadBalancer-type Kubernetes service for distributing external traffic to peer nodes evenly.

These best practices and the relevant code snippets, when presented, provide a foundational guide to setting up a resilient and highly available Hyperledger Fabric network on EKS. However, deploying such a system in a production environment requires careful planning, thorough testing, and continuous monitoring to ensure it meets the necessary performance and reliability standards.

Scaling

Configuring dynamic scalability in EKS, especially for a specific application such as Hyperledger Fabric, involves a few steps. First, we need to set up a HPA, which automatically scales the number of pod replicas in a Kubernetes Deployment or ReplicaSet based on observed CPU utilization or other select metrics. The following script demonstrates how to configure HPA for a Kubernetes deployment, which could be part of a Hyperledger Fabric network:

```
apiVersion: autoscaling/v2beta2
kind: HorizontalPodAutoscaler
metadata:
  name: hlf-peer-autoscaler
  namespace: hyperledger
spec:
  scaleTargetRef:
    apiVersion: apps/v1
    kind: Deployment
    name: hlf-peer
  minReplicas: 1
  maxReplicas: 5
  metrics:
  - type: Resource
    resource:
      name: cpu
      target:
        type: Utilization
        averageUtilization: 50
```

This script sets up an HPA named `hlf-peer-autoscaler` that targets a deployment named `hlf-peer`. It specifies that the number of replicas should be between 1 and 5, scaling up when the average CPU utilization exceeds 50%.

Adjust the `minReplicas`, `maxReplicas`, and `averageUtilization` values based on the specific needs of the solution and performance observations. It's also important that the pods' resource requests (and limits, optionally) are set in their deployment configuration, as HPA uses these to make scaling decisions. If you have custom metrics or more sophisticated scaling requirements, you might need to configure additional metric sources and modify the HPA configuration accordingly.

Save the above YAML script to a file, e.g., `hlf-peer-autoscaler.yaml`, and apply the configuration using **kubectl**:

```
kubectl apply -f hlf-peer-autoscaler.yaml
```

To monitor the status and see if the scaling is working, use the following:

```
kubectl get hpa -n hyperledger
```

This script serves as a starting point. In a real-world application, especially for an enterprise-grade solution such as Hyperledger Fabric on EKS, we would likely need a more complex setup, including fine-tuned resource requests, sophisticated metric monitoring, and perhaps additional tools for observability and management.

The next section completes this introduction to Hyperledger Fabric on EKS by looking at some quick practices for testing and troubleshooting the deployed platform.

Testing and troubleshooting a Hyperledger Fabric blockchain network on EKS

Testing and troubleshooting a Hyperledger Fabric blockchain network deployed on EKS is a crucial step to ensure its reliability, performance, and security. Effective testing strategies include deploying chaincode in a development environment, performing load testing to understand the network's behavior under stress, and conducting security vulnerability scans to identify potential threats.

After deploying the Hyperledger Fabric network on EKS, test network functionality by invoking and querying chaincode and monitoring the network for potential issues. Use the following tools and techniques for troubleshooting:

- Inspect network component logs using `kubectl logs <pod-name> -n hyperledger-fabric`

- View detailed resource information and identify issues or misconfigurations with `kubectl describe <resource-type> <resource-name> -n hyperledger-fabric`

- Execute commands inside a pod for further inspection and debugging using `kubectl exec -it <pod-name> -n hyperledger-fabric -- <command>`

Consider these solutions for common issues:

- If a pod remains in the **Pending** state, check pod events and logs for issues such as insufficient resources or scheduling constraints

- If a pod frequently restarts or crashes, examine the logs for error messages and stack traces that may indicate the cause of the problem

- If network components cannot communicate, verify the network policies and security group settings to ensure the required ports are open and the components can communicate with each other

Tools such as **Caliper**, a blockchain benchmarking tool, can be used to measure the performance of the Hyperledger Fabric network by executing defined use cases. For troubleshooting, leveraging EKS logging and monitoring services, such as **Amazon CloudWatch**, provides insights into the network's operation, enabling the quick identification and resolution of issues. Examining pod logs, monitoring resource utilization, and using Kubernetes commands such as `kubectl logs` or `kubectl describe pod` are fundamental practices for diagnosing problems. Additionally, implementing a CI/CD pipeline for automated testing and deployment can help catch issues early in the development cycle. Regularly updating the Hyperledger Fabric network components and keeping abreast of the latest security patches are also essential for maintaining a secure and efficient blockchain network on EKS.

Summary

This chapter focused on the integration of AWS's EKS and Hyperledger Fabric for blockchain hosting. We detailed the creation of an EKS cluster via the AWS Management Console and kubectl configuration, as well as the deployment of Hyperledger Fabric's components on EKS. We also looked at best practices and tools for enhancing the security, high availability, and scalability of the deployed platform. The last section was dedicated to a few strategies and tools for testing and troubleshooting infrastructure and software components.

In the next chapter, we will dive into Amazon Quantum Ledger Database, a fully managed ledger database that offers a transparent, immutable, and cryptographically verifiable transaction log owned by a central trusted authority.

Further reading

- Amazon EKS:

  `https://docs.aws.amazon.com/eks/`

- Hyperledger Fabric:

  `https://hyperledger-fabric.readthedocs.io/`

6

Building Records with Amazon Quantum Ledger Database

In this chapter, we will explore the powerful capabilities of Amazon **Quantum Ledger Database** (**QLDB**), a fully managed ledger database provided by **Amazon Web Services** (**AWS**), and how it can be utilized in the context of blockchain technology. We will cover the following topics:

- Introduction to Amazon Quantum Ledger Database

- Creating a QLDB instance

- Data modeling on QLDB

- Querying data on QLDB

- Key considerations for security, scalability, and monitoring

Introduction to Amazon Quantum Ledger Database

Amazon QLDB is a fully managed ledger database provided by AWS that offers a transparent, immutable, and cryptographically verifiable transaction log for blockchain applications. QLDB is designed to provide a high-performance and scalable solution for storing and querying transactions in a tamper-evident and tamper-resistant manner. It eliminates the need for setting up and maintaining a traditional blockchain infrastructure and can be used as a reliable data source for decentralized applications.

In this section, we will learn about the key features and benefits of using QLDB for blockchain applications, such as the following:

- **Immutability**: QLDB stores all transactions in a tamper-evident manner, ensuring that no one can alter the transaction history

- **Cryptographically verifiable**: All transactions in QLDB are cryptographically chained and hashed, allowing for easy verification of the data's integrity

- **Scalability**: QLDB is designed to scale automatically with the application's requirements, allowing organizations to handle increasing transaction loads without any manual intervention

- **Security**: QLDB is integrated with AWS **Identity and Access Management (IAM)**, providing fine-grained access control and secure communication between the QLDB instance and software applications

- **Easy integration**: QLDB provides a simple API for querying and manipulating data, making it easy to integrate with existing blockchain applications

One more feature of QLDB is its **streaming capability**. Streaming refers to a near-real-time flow of any changes to data in QLDB, which are managed via Amazon Kinesis Data Streams. The architecture design in the next section shows the building blocks of QLDB and the integration with Kinesis Data Streams.

Architecture

QLDB is designed with a unique architecture that combines elements of traditional database systems with the immutability and verifiability features of blockchain technology. The following figure represents the building block of the QLDB architecture:

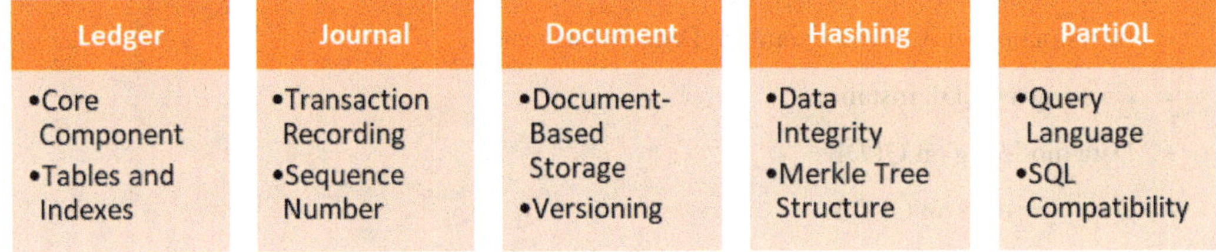

Figure 6.1 – Building blocks of Amazon QLDB

Component by component, we can identify the following characteristics:

- **Ledger**: The ledger in QLDB is the central component. It's akin to a database in traditional database systems but with added characteristics of immutability and transparency. Within each ledger, we can create multiple tables to store structured data. We can also create indexes on these tables to optimize query performance.

- **Journal**: The journal represents an immutable transaction log. All changes (inserts, updates, and deletes) are recorded in a transaction log. This log is append-only, meaning data can be added but not modified or deleted, ensuring the immutability of the ledger. Each transaction in the journal is assigned a unique, monotonically increasing sequence number, which helps in maintaining the chronological order of transactions.

- **Document**: QLDB stores data in a document-oriented format using Amazon Ion, a superset of JSON. Each time a document is updated, QLDB creates a new version of that document, while retaining all previous versions. This versioning mechanism is key for tracking the history and changes of each document over time.

- **Cryptographic hashing**: QLDB uses SHA-256 cryptographic hashing to chain document revisions and transactions together, providing a secure and tamper-evident history. The hashes are organized in a Merkle tree, a binary tree of hashes, which allows for efficient and secure verification of the data.

- **PartiQL**: QLDB supports PartiQL, a SQL-compatible query language, which allows for traditional SQL-style interaction with the data, including `INSERT`, `SELECT`, `UPDATE`, and `DELETE` operations.

> **QLDB architecture**
>
> This architecture allows QLDB to combine the flexibility and ease of use of traditional database systems with the immutable and verifiable nature of a ledger database, making it suitable for applications that require a complete and verifiable history of all data changes.

In addition to these core components, access to QLDB is managed through AWS IAM, allowing for fine-grained control over who can access and perform operations on the ledger. Also, from a network security perspective, data in QLDB is encrypted at rest and in transit. Network security is managed through AWS **Virtual Private Cloud** (**VPC**) and other AWS networking services.

Lastly, the AWS Management Console provides a user interface for managing QLDB ledgers, including creating ledgers, tables, and indexes, and executing queries.

QLDB and blockchain

Amazon QLDB offers a unique proposition for blockchain-like applications by providing an immutable, transparent, and verifiable transaction log, albeit without the decentralized consensus mechanism typical of blockchain networks. QLDB is designed for use cases where a centralized, trusted authority is acceptable, and it excels in scenarios requiring the integrity, auditability, and history tracking of data.

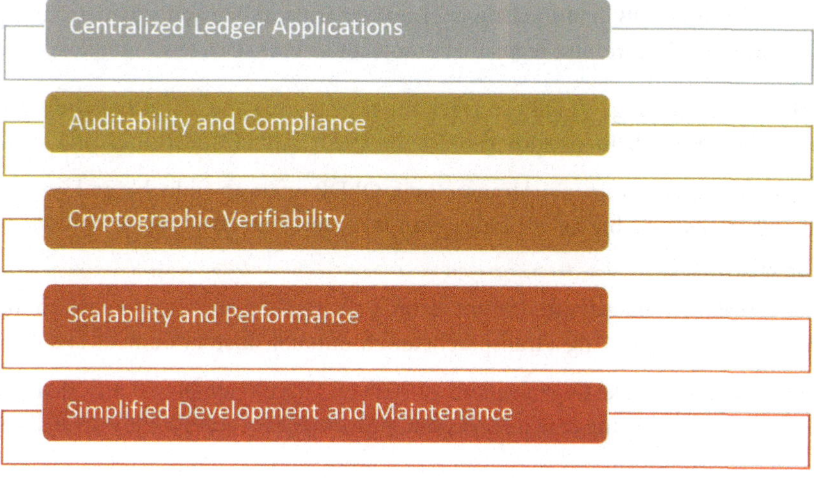

Figure 6.2 – Uses of QLDB for blockchain applications

QLDB serves as a centralized ledger that is managed by a single entity, making it suitable for organizations that need to maintain a tamper-evident history of their data but do not require a decentralized trust model (**Centralized Ledger Applications**). This is particularly useful in finance, supply chain, healthcare, and regulatory compliance where the integrity of transaction history is paramount.

With its immutable transaction log, QLDB enables straightforward auditability of all changes, making it ideal for applications where compliance and regulatory oversight are important. Auditors can easily verify the history and integrity of the data without relying on complex external systems or third-party verification (**Auditability and Compliance**).

QLDB uses cryptographic hashing to create a secure and immutable record of transactions. Each record's hash is chained to the next, creating a sequence that can be verified cryptographically (**Cryptographic Verifiability**). This ensures that any attempt to alter the history of transactions can be detected, providing a level of data integrity similar to blockchain technologies.

Unlike traditional blockchain technologies, which can suffer from slow transaction speeds and scalability issues due to the consensus mechanisms required in a decentralized network, QLDB is designed for high performance and scalability. It can execute two to three times more transactions per second than common blockchain frameworks, making it suitable for enterprise-grade applications (**Scalability and Performance**).

Developing applications on QLDB is simpler than on a decentralized blockchain platform because there is no need to implement a consensus mechanism or manage a network of nodes (**Simplified Development and Maintenance**). This can lead to lower development and operational costs, as well as easier maintenance and scaling of the application.

Use cases for blockchain

The use cases for using QLDB in blockchain applications are multiple and varied, from supply chain to financial systems, healthcare, and regulatory compliance. Let's see a few examples:

- **Supply chain tracking**: QLDB can track the provenance and status of goods as they move through the supply chain, providing an immutable record of transactions and changes

- **Financial systems**: For financial transactions, QLDB offers an auditable and verifiable history of all operations, suitable for banking, insurance, and stock trading systems where integrity is crucial

- **Healthcare data management**: Patient records, treatment histories, and consent forms can be managed in QLDB, offering a transparent and immutable history of medical data

- **Regulatory compliance**: Organizations can use QLDB to maintain records that must comply with regulations, ensuring data integrity and providing easy access to historical data for audits

> **Smart contracts**
>
> While QLDB does not natively support smart contracts in the way blockchain platforms such as Ethereum do, business logic can be implemented in the application layer. This allows for the execution of complex transactions and automated workflows, albeit under the control of the centralized authority managing the QLDB ledger.

For organizations looking for the benefits of blockchain in terms of data integrity, auditability, and transparency, but without the need for a decentralized control model, Amazon QLDB presents a powerful alternative. It combines the security features of blockchain with the scalability, ease of use, and performance of a managed AWS service, making it an attractive option for a wide range of applications.

> **QLDB versus blockchain on data privacy**
>
> When evaluating technologies such as QLDB and blockchain for data management needs, it's crucial to consider how ledger settings, particularly encryption and data retention options, impact compliance and data governance. These factors play a significant role in ensuring that data management practices align with regulatory requirements, protect sensitive information, and enable effective data governance.
>
> Amazon QLDB provides a centralized ledger with built-in encryption and the ability to manage data retention through the deletion of ledger history if needed. This can simplify compliance with regulations that require data deletion or anonymization after a certain period.
>
> Blockchain's immutable nature means that once data is added to the blockchain, it cannot be altered or deleted, which might raise concerns for compliance with regulations requiring the right to be forgotten, as stipulated by GDPR, for example. Private or permissioned blockchains may offer mechanisms to address these concerns but require careful planning and implementation.

Creating a QLDB instance

Time to move on to putting into practice what we've learned about QLDB in the previous sections. In this section, we'll create a QLDB instance. The process involves the following steps:

1. Sign in to the AWS Management Console and navigate to the Amazon QLDB console.

2. Click on **Create ledger** and provide a name for your ledger.

3. Choose the desired settings for your ledger, such as encryption, performance, and data retention options.

4. Review your settings and click on **Create** to provision your QLDB instance.

5. Once your QLDB instance is created, you can configure the necessary IAM policies and roles to provide secure access to your application.

Troubleshooting

When setting up QLDB for the first time, we may encounter initial setup issues. Some common troubleshooting tips to help address these problems include the following:

- **Check IAM permissions**: Insufficient permissions can prevent you from creating or accessing a QLDB ledger:

  Solution: Ensure that the IAM user or role you're using has the necessary permissions for QLDB. This includes permissions to create, view, and manage ledgers. Amazon provides pre-defined IAM policies for QLDB, such as `AmazonQLDBFullAccess`, which you can attach to your IAM user or role.

- **Verify service limits**: Exceeding service limits can result in errors when attempting to create new ledgers or resources:

  Solution: Check the AWS service limits for QLDB in your account and region. If you're approaching or exceeding these limits, you can request an increase through the AWS Service Quotas console.

- **Confirm region availability**: QLDB may not be available in all AWS regions:

  Solution: Check the AWS regional services list to ensure that QLDB is available in your selected region. If it's not available, consider using a different region that supports QLDB.

- **Use the correct AWS SDK and CLI versions**: Older versions of the AWS SDK or CLI might not support all QLDB features.

  Solution: Ensure you're using a recent version of the AWS SDK or CLI that supports QLDB. Update your SDK or CLI to the latest version if necessary.

- **Troubleshoot ledger creation failures**: Errors during ledger creation can occur for various reasons, including incorrect configurations or naming issues:

  Solution: When creating a ledger, ensure that the name is unique within your account and region and that it conforms to the naming rules for QLDB ledgers. Also, review any error messages closely for specific details about the failure.

Creating the ledger from the AWS Management Console

The following screenshot shows the **Create ledger** screen with the options to enter a name and select the ledger permissions mode. The **Standard** permissions mode is the recommended one, as it allows writing policies that grant or deny permissions at table level. The other mode, **Allow all**, doesn't allow table-specific permissions.

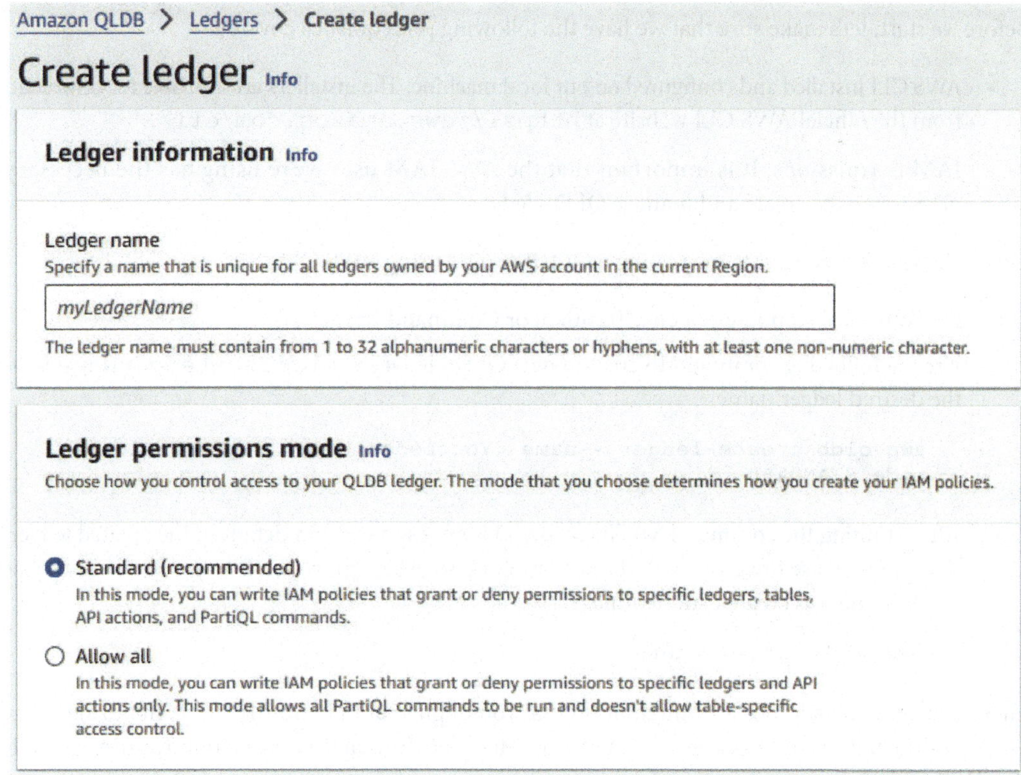

Figure 6.3 – Creating a ledger in Amazon QLDB

After creating our first instance of QLDB, we will also look at best practices for configuring the QLDB instance for optimal performance and security, using AWS **Key Management Service** (**KMS**) for encryption and configuring data retention policies.

Before doing that, though, we'll look at creating an instance of QLDB by command line. This approach will be useful for resource automation – for example, in a DevOps environment.

Using AWS Command-Line Interface

Creating a new instance of QLDB using AWS scripts typically involves using AWS **Command-Line Interface** (**CLI**) or AWS **Software Development Kits** (**SDKs**) in programming languages such as Python, JavaScript (Node.js), and so on. Based on the author's experience, the following is a general guide on how to do this using the AWS CLI, which is often the simplest method for scripting purposes.

Before we start, let's make sure that we have the following prerequisites covered:

- AWS CLI installed and configured on our local machine. The installers are available for download from the official AWS CLI website at `https://aws.amazon.com/cli/`.

- IAM permissions: It is important that the AWS IAM user we're using has the necessary permissions to create and manage QLDB ledgers.

This is the step-by-step guide to creating a QLDB Ledger using AWS CLI:

1. On your local computer, open a Terminal or Command Prompt.

2. Use the following command to create a new QLDB ledger. Replace `YourLedgerName` with the desired ledger name:

   ```
   aws qldb create-ledger --name <YourLedgerName> --permissions-
   mode STANDARD
   ```

3. After running the command, you'll receive a JSON response with details of the created ledger. You can now use the `list-ledger` command to list all your QLDB ledgers and verify that this last one was created successfully:

   ```
   aws qldb list-ledgers
   ```

There may be some additional configuration that you might want to consider for configuring other aspects of the ledger, such as encryption settings, tags, or the deletion protection feature. Consult the AWS QLDB documentation at `https://docs.aws.amazon.com/qldb/latest/developerguide/ql-db.html` for more details on these options.

Using the AWS SDK for Python

Another common approach to creating a QLDB ledger is to use the AWS SDK, available for different programming languages. The general process involves initializing an AWS client for QLDB and then calling a method similar to the CLI command's functionality. The following example is written in Python and leverages the Boto3 library.

The code snippet first initializes a QLDB client with the `boto3.client('qldb')` method, and then creates a new ledger by invoking the `create_ledger()` method on the client:

```python
import boto3

qldb_client = boto3.client('qldb')
response = qldb_client.create_ledger(
    Name='YourLedgerName',
    PermissionsMode='STANDARD'
)
```

Very simple and straightforward. Now, let's talk about data modeling.

Data modeling in QLDB

Data modeling in QLDB is an essential process that involves designing the structure and format of data to be stored in the ledger. Unlike traditional relational databases, QLDB is optimized for ledger-like transactions, making it highly suitable for tracking history and changes over time. The data model of choice impacts how users will interact with the data, the efficiency of queries, and the overall performance of an application.

The following is a summary of a few key considerations and strategies for data modeling in QLDB. First of all, let's look at **document design**, and specifically, the choice to make between a single document or multiple documents, which is deciding whether to store information in a single, comprehensive document or spread it across multiple documents. Single documents can simplify queries but may become large and unwieldy, while multiple documents can be more manageable but require more complex queries to aggregate information.

Because of (or thanks to) the document-oriented storage, and unlike relational databases that normalize data to reduce redundancy, QLDB often benefits from some level of **denormalization**. This approach can reduce the need for joining documents, improving read performance.

So, how are data relationships handled in QLDB? There are two approaches that can be taken: by reference or by embedding documents. We can **reference** other documents within a document, similar to foreign keys in relational databases. This is useful for maintaining relationships between different pieces of data. Alternatively, we can **embed related documents** within a parent document if the relationship is tightly coupled and the volume of related data is manageable.

QLDB allows us to create indexes on document fields to speed up query performance. Strategic **indexing** is crucial, especially for frequently queried fields. However, indexes must be defined at the creation time of the ledger and cannot be added or removed later, so careful planning is required.

Lastly, a word about versioning and data history. One of QLDB's key features is its immutable transaction log, which automatically versions documents upon update. This allows for easy tracking of changes over time but also means that our data model should account for the fact that historical data will be retained indefinitely.

> **Document-oriented storage**
>
> QLDB stores data in a document-oriented format using Amazon Ion, a superset of JSON that supports additional data types such as timestamps, which are crucial for ledger applications. Each record in QLDB is stored as an Ion document.

Data modeling pitfalls to avoid

When working with QLDB, effective data modeling is crucial for leveraging its full potential while ensuring efficient performance and ease of use. However, certain pitfalls in data modeling can lead to suboptimal utilization of QLDB's capabilities. The following figure summarizes common data modeling pitfalls to avoid.

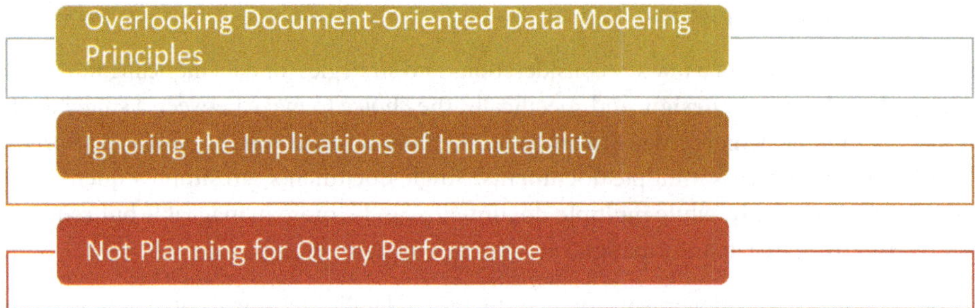

Figure 6.4 – Common data modeling pitfalls to avoid in QLDB

Let's expand on each of those pitfalls with more details.

Overlooking document-oriented data modeling principles

QLDB is designed to store data in a document-oriented format using PartiQL, a SQL-compatible query language, for interaction. A common pitfall is approaching data modeling in QLDB with a traditional relational database mindset, which can lead to inefficient data structures and queries.

The approach to take is to embrace document modeling—that is, thinking in documents and not tables. Design your data model around **documents** that can store complex nested data structures. This allows for a more natural representation of hierarchical data and can reduce the need for joins that are typical in relational models.

Discover **nested structures**. Take advantage of QLDB's support for nested objects and arrays to group related data together in a single document. This can significantly improve query efficiency by minimizing the need to access multiple documents or perform complex joins.

Don't be afraid of **denormalizing data** where appropriate. While normalization is a key principle in relational databases to reduce redundancy, QLDB's document model allows for a more denormalized approach. This can lead to simpler queries and faster access patterns. However, be mindful of the trade-off between duplication and performance, ensuring that any duplicated data does not lead to inconsistencies.

Ignoring the implications of immutability

Another pitfall is not fully considering the implications of QLDB's immutable ledger. Once data is committed to the ledger, it cannot be altered or deleted. This immutable record is a core feature for auditability and verifiability but requires careful consideration in data modeling.

A practical tip is to plan for immutability. Immutable does not mean unchangeable. While the ledger itself is immutable, the current state of your data (the latest document revision) can be changed by adding a new revision. Plan your application logic to accommodate this, especially when dealing with updates or corrections.

Also, design for auditability. Structure your documents and transactions in a way that leverages QLDB's immutability for audit purposes. This includes considering how you'll query historical data and the metadata (e.g., document versions and transaction IDs) that you'll need to retain for audit trails.

Not planning for query performance

Efficient querying is vital for any database application. A common mistake in QLDB data modeling is not considering how your data model affects query performance.

Practical tip: optimize for query patterns with proper indexing strategies and query optimization. Properly index your documents based on your query patterns. QLDB allows you to create indexes on document fields, which can drastically improve query performance. However, indexes need to be defined before any data is inserted into a table for them to be effective. Understand how PartiQL queries are executed in QLDB and optimize your queries accordingly. Avoid overly complex queries that span multiple documents or require extensive computation.

By avoiding these common pitfalls in data modeling with Amazon QLDB, you can ensure your application is well-designed, efficient, and fully capitalizes on the unique features QLDB offers.

Data model for a blockchain supply chain application

Designing a data model for a supply chain blockchain application using QLDB involves creating a structure that captures all essential elements of the supply chain, such as products, transactions, participants, and events. This example data model in the next figure, along with the AWS CLI instructions, aims to provide a foundation that can be customized for specific supply chain needs:

Products	Participants	Transactions	Events
• Product ID	• Type	• Date	• Date
• Name	• Contact	• Quantity	• Type
• Description	• Email	• Product	• Product
		• Participant	• Participant

Figure 6.5 – Example data model for a blockchain supply chain application

The data model includes the following tables:

- **Products**: Stores information about the products, including product ID, name, description, and other relevant details

- **Participants**: Records details about participants in the supply chain, such as manufacturers, distributors, retailers, and customers

- **Transactions**: Logs transactions involving products, including transfers, sales, and purchases

- **Events**: Captures events in the product life cycle, such as manufacturing, shipping, and receiving

We don't have to specify an additional table for storing historical data, as QLDB automatically maintains an immutable history of all changes, making it an ideal platform for supply chain applications where tracking the provenance and history of products is crucial.

Don't forget that this example provides a basic framework for a supply chain application on QLDB. Depending on the specific requirements of a more sophisticated application, we may need to add more tables, fields, and indexes, or adjust the document structure to best fit the identified use case.

Best practices for data modeling

Data modeling in QLDB requires a different mindset than traditional relational databases, emphasizing document design, denormalization, and the efficient use of indexes. By carefully considering these factors, we can leverage QLDB's strengths to build robust, scalable applications that benefit from its ledger capabilities and historical data tracking. The following figure summarizes four typical best practices for an optimal data model:

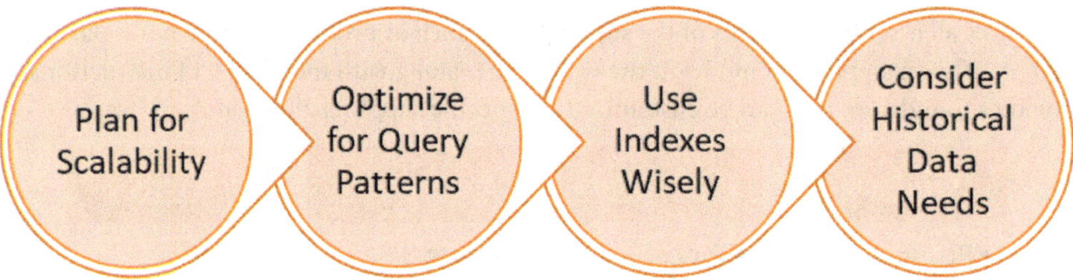

Figure 6.6 – Best practices for data modeling in QLDB

Let's expand on each practice:

- **Plan for scalability** refers to designing the data model with future growth in mind, considering both the size of individual documents and the overall data set.

- **Optimize for query patterns** is all about understanding an application's query patterns and modeling its data to facilitate efficient querying. This may involve denormalizing data or strategically splitting information across documents.

- **Use indexes wisely** means creating indexes on fields that will be frequently used as query predicates, but be mindful of the limitations and the fact that they cannot be altered after the ledger's creation.

- **Consider historical data needs** implies designing the model considering how the application will query historical data, leveraging QLDB's immutable history to track changes over time without additional complexity in the application logic.

We'll make a few tangible examples of each best practice in the next sections. Now, let's proceed with the creation of a document inside the instance of the ledger previously created.

Creating a document

To create a document in QLDB, we use the PartiQL query language, which is SQL-compatible and allows us to interact with our ledger data. Documents in QLDB are stored in Ion format, which is similar to JSON but with additional data types and features suitable for ledger data.

The basic command structure to insert a document into a table in QLDB using PartiQL is as follows:

```
INSERT INTO <TableName> << { 'field1': 'value1', 'field2': 'value2',
... } >>
```

Assuming we have a table named `Customers`, we want to insert a document with the name and additional information of a customer. The `INSERT` command would look something like this:

```
INSERT INTO Customers << { 'name': 'John Doe', 'email': 'john.doe@
example.com', 'isActive': true } >>
```

To run this command via the AWS CLI, we would wrap it in a call to the `execute-statement` command, specifying the ledger name and the PartiQL statement, like so:

```
aws qldb execute-statement --ledger-name YourLedgerName --statement
"INSERT INTO Customers << { 'name': 'John Doe', 'email': 'john.doe@
example.com', 'isActive': true } >>"
```

The actual AWS CLI command might differ since the `execute-statement` command is not directly available in the AWS CLI for QLDB. Typically, we would execute PartiQL statements through a programming language SDK (e.g., Python, Java, or Node.js) or the QLDB Query Editor in the AWS Management Console.

The preceding command is conceptual. In practice, we would use the AWS SDKs such as Boto3 for Python, the AWS SDK for Java, or another SDK to execute PartiQL commands against QLDB. Each SDK has its own method for executing a statement, which involves creating a session with the ledger and then executing the statement within that session.

Here's an example using Python with Boto3. After creating an instance of a client object to a QLDB session, the `execute_statement` function executes the PartiQL command:

```
client = boto3.client('qldb-session')
def execute_statement(ledgerName, statement):
    response = client.send_command(
        SessionToken=<YourSessionToken>,
        Statement=statement,
        LedgerName=ledgerName
    )
    return response
```

In the preceding example, `YourSessionToken` is a placeholder for a session token that we would obtain by starting a session with the QLDB ledger. Managing sessions and executing statements typically involve more detailed logic to handle sessions, tokens, and parsing responses, which is why using an AWS SDK is recommended for interacting with QLDB programmatically.

Referencing and embedding documents

As mentioned before, we can structure our data in QLDB using references to other documents (similar to foreign keys in relational databases) or by embedding documents directly within other documents. This flexibility allows us to model complex relationships and hierarchies in a way that best suits any application's needs. Let's give an example of both approaches.

Consider a simple database with two tables, `Customers` and `Orders`. Each order is associated with a customer, but instead of embedding customer details in each order, we reference the `CustomerId` column in the `Orders` table. This is an example of **referencing another document**.

The `Customers` table is described by the following JSON structure:

```
{
    "CustomerId": "C1",
    "Name": "John Doe",
    "Email": "john.doe@example.com"
}
```

The `Orders` table, with the reference, will look like the following:

```
{
    "OrderId": "O1",
    "OrderDate": "2024-01-01",
    "Amount": 150.00,
    "CustomerId": "C1"
}
```

The `C1` value references the `CustomerId` column in the `Customers` table.

To create these documents in QLDB, we would first insert the customer and order records into their respective tables, ensuring that the `CustomerId` value in the `Orders` table matches the `CustomerId` value of the corresponding customer in the `Customers` table.

Let's now look at the second scenario. Now, instead of referencing the customer in each order, we embed the customer details directly within the order document. This is an example of **embedded documents**. This approach is useful when we want to snapshot the customer information at the time of the order because customer details can change over time.

The `Orders` table with the embedded customer details will have the following definition:

```
{
   "OrderId": "O1",
   "OrderDate": "2024-01-01",
   "Amount": 150.00,
   "Customer": {
     "CustomerId": "C1",
     "Name": "John Doe",
     "Email": "john.doe@example.com"
   }
}
```

In this approach, the order document contains all relevant customer information, eliminating the need to join documents to retrieve the full order and customer details.

Here are some considerations. The referencing documents approach makes it easier to maintain consistent data, especially for entities that change infrequently or where we want to maintain a single source of truth (e.g., customer profiles). However, it requires additional queries to resolve references.

Embedding provides a snapshot of related data at a specific point in time, which can be beneficial for historical accuracy and query performance since all data is contained within a single document. The trade-off is potential data duplication and the need to update embedded data in multiple places if it changes.

Choosing between these approaches depends on an application's specific requirements, such as query patterns, data update frequency, and the importance of historical accuracy versus query efficiency.

Indexing

Indexing in QLDB is crucial for improving query performance. Unlike traditional databases, where we might create indexes after populating data, in QLDB, we must define indexes at the time of table creation or before inserting any data into the tables.

To create an index in QLDB, we use the `CREATE INDEX` statement in PartiQL. This statement specifies the table and the document field (or fields) we want to index. Here's the syntax:

```
CREATE INDEX ON <TableName> (<FieldName>)
```

For example, for our table named `Customers`, we want to create an index on the `Email` field. The command would be as follows:

```
CREATE INDEX ON Customers (Email)
```

As mentioned, indexes in QLDB must be created before inserting documents into the table. This is because QLDB indexes are immutable—once set, they cannot be modified or deleted. The best practice is to choose fields that are frequently used in query predicates. Indexing these fields can significantly improve the performance of `SELECT` queries that filter or sort based on these fields.

QLDB also supports indexing multiple fields together, known as **composite indexes**. However, when querying, we need to filter by all the fields in the composite index to leverage its benefits.

While indexes improve query performance, they can slightly slow down data insertion because the index must be updated with each new entry. It's important to balance the need for query performance with the potential impact on write operations.

Versioning

QLDB inherently manages versioning and data history as part of its core functionality, providing an immutable and verifiable history of all changes over time. This feature allows users to track each document's entire history, including inserts, updates, and deletions. Every document is versioned automatically. Each time a document is updated or deleted, QLDB preserves the previous version, enabling a complete and verifiable history of changes. This feature is fundamental for use cases requiring auditability, such as financial transactions, supply chain management, and regulatory compliance.

To query the history of a document or documents within a table, we use the `HISTORY` function in PartiQL. This function allows us to retrieve the history of changes to a document, including the version metadata. The PartiQL syntax goes as follows:

```
SELECT * FROM HISTORY(<TableName>, <StartTime>, <EndTime>)
WHERE metadata.id = '<DocumentId>'
```

Here, `<TableName>` is the name of the table containing the document, and `<StartTime>` and `<EndTime>` are optional timestamp parameters to narrow down the history to a specific time range. If omitted, QLDB returns the entire history. `<DocumentId>` is the unique ID of the document whose history we want to query.

Let's have an example of retrieving the history of a document in the supply chain application, specifically from the `Transactions` table. Our query might look like this:

```
SELECT * FROM HISTORY(Transactions, timestamp '2023-01-01T00:00:00Z',
timestamp '2023-12-31T23:59:59Z')
WHERE metadata.id = '1234567890abcdef'
```

This query retrieves the history of the specified transaction for the entire year of 2023.

Managing data history

While QLDB automatically manages versioning and history, understanding how to interact with this data is crucial for managing the ledger effectively. For **auditing and compliance**, using the `HISTORY` function to audit transactions and changes will help ensure compliance with regulatory standards. From a **data integrity** perspective, QLDB's immutable history helps maintain data integrity, as it provides a transparent and verifiable record of all changes. Data history also helps **query optimization**: when querying historical data, we can be specific with our time range and conditions to optimize performance, especially when working with large datasets.

There are some limitations to consider:

- **Immutable nature**: The immutable nature of QLDB means we cannot delete history. If sensitive data must be removed, we should instead consider best practices for data handling, such as not storing direct **Personally Identifiable Information** (**PII**) or using logical deletion flags in our document design.

- **Storage impact**: Since QLDB retains an immutable history of changes, we should consider the impact on storage, especially for ledgers with high transaction volumes or large documents.

- **Query performance**: Querying historical data can be resource-intensive. The best practice is always to use precise queries to minimize performance impact.

QLDB's approach to versioning and data history is a powerful feature for applications requiring a transparent and tamper-evident history of data changes. By leveraging the `HISTORY` function and understanding how to efficiently query historical data, we can harness this capability for auditing, data analysis, and compliance purposes.

Querying data in QLDB

QLDB provides a powerful and flexible API for querying and manipulating data stored in a ledger. In this section, we will have a look at how to perform the following operations:

1. **Retrieve data**: Query data from tables using PartiQL, a SQL-compatible query language for Amazon Ion.

2. **Update data**: Use the QLDB API to update documents, ensuring that all updates are recorded as new transactions in the ledger.

3. **Verify data**: Use QLDB's built-in cryptographic features to ensure the integrity and authenticity of data by validating the cryptographic hashes and digital signatures associated with your transactions.

4. **Perform advanced queries**: Leverage QLDB's support for PartiQL to perform complex queries, such as aggregations, joins, and filtering.

At the end of this section, we will also get familiar with common query access patterns, such as using parameterized queries to prevent injection attacks and optimizing our queries for the specific access patterns of our application.

CRUD operations

Querying data in QLDB involves using PartiQL. PartiQL provides the flexibility to run SELECT, INSERT, UPDATE, DELETE, and HISTORY queries on ledger data. Let's go case by case and perform some common data querying operations in QLDB.

To retrieve data from a table in QLDB, we use the SELECT statement, similar to SQL. We can select specific fields, use conditions, and perform joins across tables. The following example retrieves all records from the Customers table:

```
SELECT * FROM Customers
```

The use of the star (*) wildcard is not recommended, though. Although it is a quick shortcut to indicate all fields in a table, the best practice is to indicate specific fields in a table:

```
SELECT CustomerId, Name, Email FROM Customers WHERE IsActive = true
```

To add a new document to a table, we use the INSERT INTO statement followed by the table name and the document content in Ion format. In this example, we'll insert a new customer named Jane Doe into the Customers table:

```
INSERT INTO Customers << {
    'Name': 'Jane Doe',
    'Email': 'jane.doe@example.com',
    'IsActive': true
} >>
```

To modify existing documents, we'll use the UPDATE statement, specifying the table, the set of changes, and a condition to select the document(s) to update. For example, let's update the email address for the customer identified by having CustomerId = 12345:

```
UPDATE Customers SET Email = 'new.email@example.com' WHERE CustomerId
= '12345'
```

Lastly, to remove documents from a table, we can use the DELETE FROM statement with a condition to specify which documents to delete. Again, as an example, let's delete a customer from the Customers table:

```
DELETE FROM Customers WHERE CustomerId = '12345'
```

Executing queries programmatically

While we can execute all these queries directly in the AWS Management Console using the QLDB Query Editor, we may also want to run them programmatically using AWS SDKs, such as Boto3 for Python, AWS SDK for JavaScript, and so on. The following is a Python example that initializes the QLDB client and defines a SELECT PartiQL query to retrieve all active customers:

```
client = boto3.client('qldb')
query = "SELECT * FROM Customers WHERE IsActive = true"

response = client.execute_statement(
    LedgerName=ledger_name,
    Statement=query
)

print(response)
```

This example is a simplified illustration. In practice, executing a query involves handling sessions and possibly pagination for the query results.

> **Note**
>
> QLDB's execution model is different from traditional databases, and managing sessions, transactions, and query results can vary significantly based on the SDK and programming language in use. Always refer to the specific AWS SDK documentation for detailed examples and best practices for managing database interactions.

Data verification

QLDB provides built-in cryptographic features to ensure the integrity and verification of the data stored within it. This is achieved through the use of an immutable transaction log, cryptographic hashing, and a concept known as Merkle trees. These features enable us to verify the completeness and accuracy of our data at any point in time. Let's first have an overview of how data verification works in QLDB before diving into how we can perform it.

The cryptographic features of QLDB include the following:

- **Immutable transaction log**: Every transaction in QLDB is recorded in an append-only transaction log, ensuring that once data is written, it cannot be altered or deleted.

- **Cryptographic hashing**: QLDB uses SHA-256 cryptographic hashing to generate a unique hash for every document revision and transaction. These hashes are used to chain revisions and transactions together, providing a tamper-evident history.

- **Merkle trees**: QLDB utilizes Merkle trees to efficiently prove the integrity and completeness of the data. A Merkle tree is a binary tree of hashes, where every leaf node is a hash of transaction data, and every non-leaf node is a hash of its child nodes. The root of this tree (the Merkle root) represents the entire dataset's fingerprint at a specific point in time.

Performing data verification involves obtaining the digest of a ledger for a specific point in time and then using that digest to verify the integrity of the data. The following instructions will guide us in performing these steps:

1. **Obtain the ledger's digest**: A ledger's digest provides a secure output (hash) representing the entire state of the ledger at a specific point in time. We can obtain the digest using AWS CLI with the `get-digest` command:

   ```
   aws qldb get-digest --name YourLedgerName
   ```

   This command returns the digest, consisting of the digest's hash and a block address (sequence number and hash) marking the ledger's state.

2. **Verify a document or transaction**: After obtaining the ledger's digest, we can verify a particular document or transaction against this digest. This involves retrieving a document's revision history or a transaction block, and then requesting proof that this revision or block is included in the digest, by using the `get-revision` command:

   ```
   aws qldb get-revision --name YourLedgerName --block-address
   '{\"IonText\": \"{\\\"strandId\\\": \\\"yourStrandIdHere\\\",
   \\\"sequenceNo\\\": yourSequenceNoHere}\"}' --document-id
   "yourDocumentIdHere"
   ```

   With this command, first, we get a revision at a specified block address – that is, the command retrieves the specific document revision or transaction block that we want to verify. This will give us the data and its associated metadata, including the hash.

   Then, we request proof for the document revision or transaction block. The proof is a list of hashes that, when combined with the revision or block's hash, should match the ledger's digest.

   The command then uses the proof and the revision or block's hash to calculate the Merkle tree. If the calculation matches the ledger's digest, it confirms that the document or transaction was indeed part of the ledger at the time the digest was obtained.

There are some considerations to keep in mind when performing data verification. Frequent data verification operations can incur costs, so it's essential to understand the pricing model of QLDB and optimize the verification processes accordingly. Ideally, we want to automate the verification process for applications requiring regular verification of data integrity. Automating the data verification process involves scheduled checks or triggers based on specific events and the use of the AWS SDK for QLDB.

Common query access patterns

When working with QLDB, optimizing data models and queries based on common access patterns can significantly enhance performance and efficiency. Understanding and planning for these patterns is crucial for designing optimal application interactions with QLDB. Let's look at some common query access patterns and strategies, as depicted in the following table:

Optimization	Pattern	Strategy
Direct Lookup by Unique Identifier	Fetching a document or set of documents using a unique identifier, such as a customer ID, transaction ID, or any other unique key	Use an indexed field for the unique identifier to ensure efficient retrieval. Ensure you create an index on any field used as a unique identifier to optimize these lookups.
Historical Data Retrieval	Querying the history of a document to track changes over time, which is a distinctive feature of QLDB	Utilize the `HISTORY` function to retrieve the history of a document based on its ID. This access pattern is inherently optimized by QLDB's design, but understanding how to filter and limit results can help manage performance and data volume.
Conditional Queries	Searching for documents based on specific conditions, such as date ranges, status flags, or partial matches on strings	Create indexes on fields that are frequently used in query conditions to improve performance. Be mindful that QLDB indexes need to be set up before inserting data.
Aggregation Queries	Summarizing or aggregating data, such as counting documents, summing values, or averaging	While QLDB supports SQL-like aggregation functions (`COUNT`, `SUM`, `AVG`, etc.), consider the volume of data being aggregated and the potential impact on performance. For large datasets, consider strategies to limit the scope of aggregation.
Joining Documents	Combining data from multiple documents or tables, such as joining customer information with their orders	Although QLDB supports JOIN operations, they can be resource-intensive. Design your data model to minimize the need for joins, possibly by denormalizing data where appropriate. When joins are necessary, ensure that the fields being joined are indexed.

Pagination and Batch Retrieval	Retrieving large sets of data in manageable chunks, especially for applications with large datasets or that serve data to frontend interfaces	Implement pagination in your application logic. QLDB does not natively support SQL-style `LIMIT` and `OFFSET` keywords, so pagination typically involves application-side logic to manage data retrieval in batches.
Audit and Compliance Checks	Performing queries to support audit and compliance requirements, which may involve complex queries across the ledger's history	Leverage the immutability and cryptographic verification features of QLDB to perform these checks. Efficiently structure these queries to minimize performance impacts, and consider offloading intensive audit operations to off-peak hours if possible.

Table 6.1 – Common query access patterns and strategies

By understanding these common access patterns and adopting the corresponding strategies, we can design QLDB applications to be more efficient, responsive, and scalable. Additional considerations for the optimization of a QLDB instance are also presented in the next section.

Key considerations for security, scalability, and monitoring

When using QLDB, it is essential to consider several factors to ensure the security, scalability, and monitoring of solutions that rely on data stored in QLDB. In this section, we will present key considerations to adopt for the following:

- **Security**: How to secure a QLDB instance by configuring IAM policies and roles, enabling encryption at rest using AWS KMS, and using secure communication channels such as HTTPS and VPC endpoints

- **Scaling**: How to design the data model and queries to optimize performance and manage the load on a QLDB instance, allowing it to scale automatically with the application's requirements

- **Monitoring**: How to monitor a QLDB instance using AWS CloudWatch and the QLDB console to track performance metrics, identify issues, and take appropriate action when needed

By implementing the following best practices, we can ensure that an application based on QLDB remains secure, scalable, and easy to manage.

Security

When using QLDB, ensuring the security of data and transactions is paramount. AWS provides robust tools and features to help secure QLDB ledgers, but it's also crucial to follow best practices and understand the shared responsibility model for cloud security. Let's start with key considerations for IAM:

- **Use IAM policies**: Define IAM policies to control access to QLDB resources. The key here is to be specific about who can perform actions such as creating ledgers, inserting data, and querying data.

- **Principle of least privilege**: Apply the principle of least privilege by granting only the necessary permissions required to perform a task. For example, if a user only needs to query data, they should not have permission to delete the ledger.

- **Use IAM roles for applications**: When accessing QLDB from other AWS services (e.g., AWS Lambda), use IAM roles to manage permissions securely.

When talking of **encryption at rest and in transit**, QLDB automatically encrypts data at rest using AWS-managed keys in AWS KMS. It is also possible to opt for customer-managed keys for enhanced control. On top of that, we need to ensure that data is encrypted in transit by using HTTPS endpoints provided by QLDB, which secure the data with TLS.

Network security plays also a significant importance in the overall security of the solution:

- Utilize **VPC endpoints** for QLDB to allow direct connection to QLDB from an Amazon VPC, without using public internet. This enhances security by keeping traffic within the AWS network.

- Apply security groups and network **Access Control Lists** (**ACLs**) to control inbound and outbound traffic to and from cloud resources.

When talking about best practices for the design of a data model, we've already mentioned data privacy and compliance, and specifically, handling sensitive data carefully. As a rule of thumb, PII and other sensitive data, when stored in QLDB, should always be protected by privacy preservation practices, such as data minimization or confidential computing.

The last but not least level of security is at the application level. Always validate inputs: it's necessary to implement robust **input validation** in our applications to prevent injection attacks and ensure that only valid data is written to the ledger. Also, it's important to follow **secure coding practices** to protect applications from vulnerabilities. A good practice, for example, is to regularly update libraries and dependencies to mitigate known security issues.

Remember, security is not a feature, it's a design element of a software application. By integrating these security considerations into our QLDB application's design and operation, we can take full advantage of QLDB's features while maintaining a high level of security.

Scalability

Scalability is a critical factor when designing and deploying applications on QLDB. Ensuring that applications can handle growth in data volume, transaction rates, and query complexity without compromising performance requires careful planning and optimization. The following figure highlights the key considerations for scalability when working with QLDB:

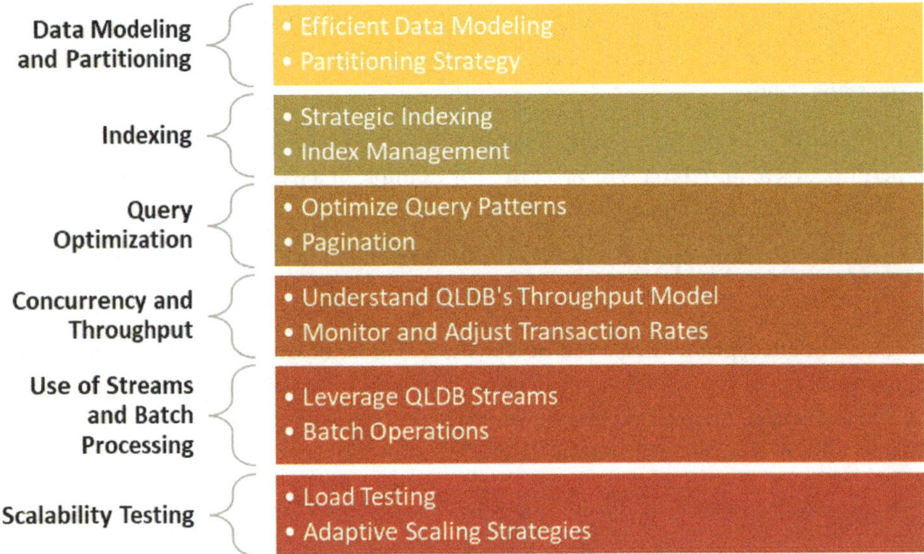

Figure 6.7 – Key considerations for scalability in QLDB

As the figure anticipates, data modeling and partitioning consist of two common approaches:

- **Efficient data modeling**: Design the data model to optimize for common access patterns. Efficiently structured data can significantly reduce the computational overhead of queries, especially as data volume grows.

- **Partitioning strategy**: While QLDB abstracts the physical data storage details, understanding how data access patterns correlate with QLDB's storage and partitioning can help in optimizing performance. For example, frequently accessed data should be modeled in a way that minimizes cross-partition queries.

Indexing is also a key factor in improving scalability, by adopting these two best practices:

- **Strategic indexing**: Create indexes on frequently queried fields to speed up query performance. However, be mindful that each index comes with a storage and maintenance cost. Indexes should be planned at the design phase, as they cannot be added or removed after data insertion.

- **Index management**: Manage indexes wisely by focusing on fields that are often used in query predicates and ensuring they are created before data ingestion.

There's no index that can make performance miracles if, at the foundation, queries are not optimized. Query optimization involves the following:

- **Optimizing query patterns**: Analyze and optimize the most common query patterns. Simplifying queries, reducing the amount of data scanned, and avoiding unnecessary joins can help maintain performance as data grows.

- **Pagination**: Implement effective pagination in an application to manage large datasets efficiently. Since QLDB does not natively support SQL-style pagination keywords, we may need to implement custom logic in the application.

Improved concurrency and increased throughput are generally positive consequences of applying the good practices described before, specifically the following:

- **Understand QLDB's throughput model**: QLDB uses an **Optimistic Concurrency Control** (OCC) model, which means it checks for transaction conflicts at commit time. Designing applications to handle retry logic for transaction conflicts is crucial for maintaining high concurrency.

- **Monitor and adjust transaction rates**: Monitor applications' transaction rates and adjust request rates accordingly. Be aware of QLDB's read and write IOPS limits and plan for scaling these limits as needed.

For more advanced capabilities, the use of streams and batch processing requires the following considerations:

- **Leverage QLDB streams**: For applications that require real-time data replication or integration with other services, consider using QLDB streams. This feature allows an application to capture document revisions in near-real-time, enabling scalable integration patterns with other AWS services.

- **Batch operations**: When possible, batch processing of transactions can reduce the overhead of writing data to QLDB. Batching is particularly effective for bulk inserts or updates.

Test, test, and test your application until you're ultimately happy. The same applies to scalability testing:

- **Load testing**: Regularly perform load testing to understand how your application behaves under high load conditions. This can help identify bottlenecks and areas for optimization before they become issues in production.

- **Adaptive scaling strategies**: Plan for scalability both horizontally (e.g., by adding more consumer applications that read from QLDB streams) and vertically (e.g., by optimizing queries and data structures).

There are many factors, but scalability in software applications is not rocket science. It requires patience and trial. By considering the presented factors from the outset and regularly reviewing applications' performance and architecture, we can ensure that our QLDB-based applications remain scalable, performant, and cost-effective as they grow.

Monitoring

Monitoring query access patterns in QLDB can help discover access data to the ledger – in particular, when, from where, and what transactions were run. This kind of ledger monitoring helps associate QLDB transactions with API requests and identify the PartiQL statement run within the transaction. It also details a solution for mapping QLDB journal blocks to corresponding entries in API logs. This approach helps in understanding access patterns and associating client requests with ledger modifications.

The **Amazon QLDB journal** is a fundamental component of the QLDB architecture, acting as an immutable transaction log. It records every change made to the data in a ledger, ensuring complete data integrity and allowing for full traceability of the transaction history. Each transaction in the journal is assigned a unique, sequential identifier, enabling the chronological tracking of changes. This design ensures that once a transaction is recorded, it cannot be altered or deleted, providing a tamper-evident history, which is crucial for applications requiring auditability and data integrity. The journal's immutability is key to the trust and reliability offered by QLDB.

An Amazon QLDB journal block typically contains a structured record of one or more transactions that have been committed to the ledger. Each block in the journal includes the following:

- **Block metadata**: This includes the unique sequence number of the block, the timestamp of when the block was written, and cryptographic hash values.

- **Transaction details**: For each transaction, the block records the transaction ID, the PartiQL statements executed, and any resulting changes to the data. This might include details of documents inserted, updated, or deleted.

- **Cryptographic hashing**: The block includes a hash of its contents, along with a hash that links it to the previous block in the chain, ensuring immutability and verifiability of the transaction history.

The content of a journal block can be described by this JSON structure:

```
{
    "JournalBlock": 12345,
    "Timestamp": "2024-02-15T10:30:00Z",
    "BlockHash": "0x91a7364...",
    "PreviousBlockHash": "0x9b5c2a1...",
    "Transactions": [
        {
            "TransactionID": "TXN-20240215-1030-00001",
            "PartiQLStatements": [
                "INSERT INTO Orders << {'OrderId': 'O1001', 'Product':
    'Widget', 'Quantity': 10, 'OrderDate': '2024-02-15'} >>"
            ],
            "DocumentRevisionMetadata": {
                "DocumentID": "some-document-id",
                "VersionNumber": "some-version-number",
```

```
        "DocumentHash": "some-hash-value"
    },
    "Changes": "New record added to the Orders table.",
    "TransactionHash": "0x3b7c8d9..."
  }
 ]
}
```

This block represents a single transaction where a new order was inserted into an `Orders` table. This JSON representation includes all the key elements of the journal block: block metadata, transaction details, PartiQL statements executed, document revision metadata, and cryptographic hashes. The structure is designed to reflect the immutable and chronological nature of transactions in QLDB.

The design of these journal blocks is a critical aspect of how QLDB maintains a tamper-evident and auditable history of all changes in the ledger. This structure provides the foundation for the ledger's integrity and trustworthiness, it includes metadata for the block, details of the transaction, and cryptographic hashes to ensure immutability and linkage to the ledger's history.

Summary

Amazon QLDB offers a unique value proposition in the realm of blockchain solutions by providing a centralized, fully managed ledger database that combines the immutable and cryptographically verifiable nature of blockchain with the ease of use, scalability, and reliability of AWS cloud infrastructure. Unlike traditional blockchains that operate on a decentralized consensus mechanism, QLDB simplifies the trust model by offering a single, authoritative data source. This approach eliminates the complexity and overhead associated with managing a decentralized network while still ensuring data integrity and transparency through an immutable transaction log.

QLDB's added value lies in its ability to seamlessly integrate with existing AWS services, offering robust data management capabilities, including fast and flexible querying with PartiQL, automatic scaling, and high availability. It is particularly well-suited for use cases where the integrity of transactional data is paramount, but the application can rely on a trusted central authority, thus bridging the gap between traditional database systems and blockchain technology. This makes QLDB an attractive solution for industries such as finance, supply chain, healthcare, and government, where verifiable and immutable record-keeping is critical, without the complexities of a decentralized blockchain architecture.

This chapter provided a comprehensive exploration of Amazon QLDB and its data model. We introduced QLDB, discussed its features, and also highlighted potential benefits and use cases for blockchain applications. We delved into data modeling and querying in QLDB, and explained how to design tables, manage data, and use the QLDB API for data retrieval and manipulation. Key considerations for security, scaling, and monitoring were also discussed to ensure optimal performance and reliability of your blockchain applications.

As we conclude this part of our journey, we look forward to the next exciting phase, where we will shift our focus to Microsoft Azure. In the upcoming chapters, we will explore hosting a Corda distributed ledger technology network on Azure Kubernetes Service, using the ledger features of Azure SQL, and leveraging Microsoft Azure confidential ledger.

Further reading

- Amazon Quantum Ledger Database:

  ```
  https://aws.amazon.com/qldb/
  ```

- Amazon QLDB Developer's Guide:

  ```
  https://docs.aws.amazon.com/qldb/latest/developerguide/what-is.html
  ```

Part 3: Deploying and Implementing Blockchain Solutions on Azure

This part covers deploying and implementing blockchain solutions on Azure, including how to host a blockchain network on Azure Kubernetes Service, using the ledger features of Azure SQL, and leveraging Microsoft Azure Confidential Ledger. Particular attention is paid to comparing the centralized ledger capabilities available in Azure with pure decentralized blockchain networks.

This part includes the following chapters:

- *Chapter 7, Hosting a Corda DLT Network on Azure Kubernetes Service*
- *Chapter 8, Using the Ledger Features of Azure SQL*
- *Chapter 9, Leveraging Azure Confidential Ledger*

7

Hosting a Corda DLT Network on Azure Kubernetes Service

In recent years, **Distributed Ledger Technology (DLT)** has emerged as a groundbreaking innovation with the potential to revolutionize industries across the globe. Among the various DLT platforms, Corda has gained significant attention for its focus on privacy, scalability, and interoperability. With its unique design tailored for enterprise use cases, Corda provides a powerful foundation for building decentralized applications and networks.

To harness the full potential of Corda, organizations require a robust and scalable infrastructure that can support the deployment and management of Corda nodes. This is where **Azure Kubernetes Service (AKS)**, a managed container orchestration service provided by Microsoft Azure, comes into play. By leveraging AKS, organizations can effectively host and manage Corda DLT networks with enhanced scalability, security, and operational efficiency.

In this chapter, we won't dive deep into the details of Corda's architecture. It would take an entire book to just describe Corda! And there's already plenty of documentation in the official channels (see the *Further reading* section at the end of this chapter).

Rather, in the next few sections, we'll delve into the intricacies of hosting a Corda DLT network on AKS. We'll explore the key considerations, best practices, and step-by-step guidelines to establish a highly available and resilient infrastructure for Corda-based applications. Whether you are an experienced Corda developer, a DevOps engineer, or a technology enthusiast looking to deepen your knowledge of DLT deployment, this chapter aims to provide you with a comprehensive understanding of AKS and its integration with Corda.

Now, let's dive into the following main topics:

- Understanding Corda and AKS
- Architecting Corda networks on AKS
- Provisioning an AKS cluster for Corda
- Managing Corda nodes on AKS

Technical requirements

To run the scripts presented in this chapter for provisioning a Corda network on AKS, you will need an Azure account with an active subscription, as well as resource administrative permissions on the Azure subscription. You can create a free Azure account at `https://azure.com/free`

This chapter assumes you're familiar with Azure and AKS for deployment tasks.

Understanding Corda and AKS

In the realm of DLT, Corda has emerged as a prominent platform with a focus on addressing the specific requirements of businesses and enterprises. Built by R3, a leading blockchain software firm, Corda is designed to facilitate secure and efficient transactions between parties while preserving data privacy and confidentiality. Its unique features make it an ideal choice for industries such as finance, supply chain, healthcare, and more. Additional information about industries and specific success case studies for Corda can be found on the R3 Corda website: `https://www.r3.com/products/corda`.

At its core, Corda is a decentralized ledger that enables participants to transact directly, eliminating the need for intermediaries and reducing transactional friction. Like the Ethereum blockchain, Corda embraces a *smart contract* approach where transactions are executed using predefined business logic known as **Corda Distributed Applications** (**CorDapps**). Unlike Ethereum's smart contracts, CorDapps are designed with a specific set of features and considerations that cater to the unique requirements of regulated financial institutions and other businesses, such as privacy, legal provisions, and a unique communication model known as *flow* to manage complex multi-step transaction protocols between parties.

Key features of Corda DLT

R3, the company that develops Corda, has published a detailed white paper that describes the architecture components of the Corda platform. The white paper is available at `https://www.corda.net/content/corda-platform-whitepaper.pdf`.

Along with relevant architecture diagrams, the white paper identifies the core components of Corda, specifically its approach to identity and consensus, the state management system, and the flow framework. We recommend that you have a look at the white paper before progressing further with this chapter.

Our focus for this chapter is to describe how Corda interacts with AKS. The following diagram summarizes the key features of a solution built on Corda as the core DLT. Corda's innovative approach to DLT, with its emphasis on privacy, scalability, interoperability, and regulatory compliance, positions it as a compelling solution for enterprises seeking to harness the benefits of blockchain technology. As we delve deeper into hosting Corda networks on AKS in this chapter, we will explore how the integration of Corda with AKS can amplify these key features and provide organizations with a robust infrastructure for their Corda-based applications:

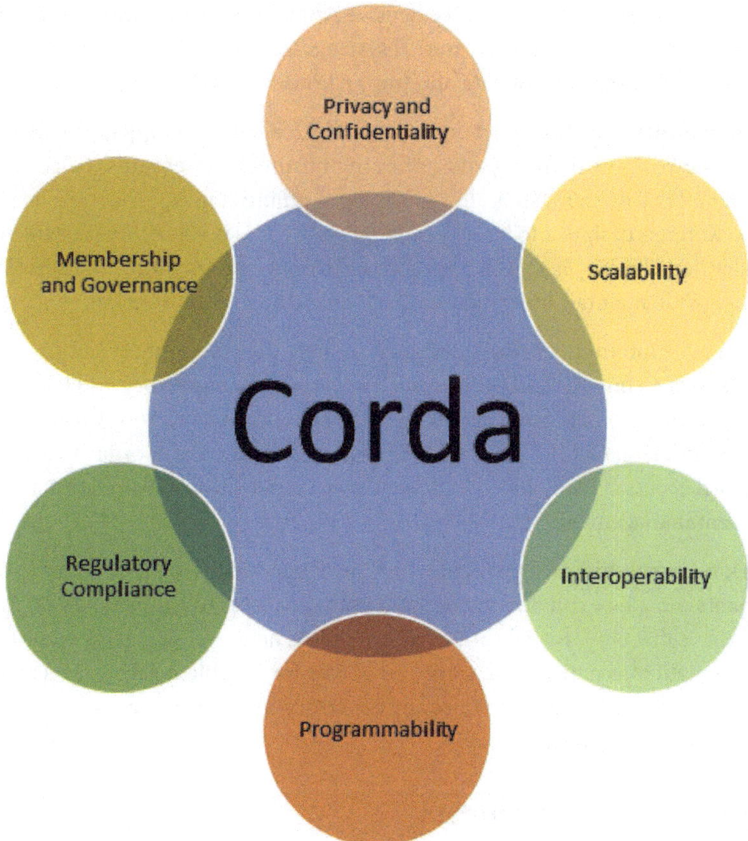

Figure 7.1 – Key features of Corda DLT

Let's expand on the key features of Corda:

- **Privacy and confidentiality**: Corda takes privacy seriously by design. Unlike public blockchains, where every transaction is visible to all participants, Corda ensures that transaction data is only shared with relevant parties. It employs an approach called **fine-grained access control** to enable selective disclosure of transaction details, allowing participants to maintain privacy while sharing information on a need-to-know basis.

- **Scalability**: Corda's architecture is optimized for scalability, allowing for high transaction throughput and network performance. Its unique approach of notarization, where only specific transactions require consensus, contributes to the efficiency of the network. Corda also supports horizontal scalability by allowing multiple parallel instances of the ledger, ensuring that the system can handle increasing transaction volumes without sacrificing performance.

- **Interoperability**: Corda recognizes the importance of interoperability within complex business networks. It enables seamless integration with existing systems and platforms through its pluggable consensus mechanism and extensive **Application Programming Interfaces (APIs)**. This allows Corda to interact with external systems, legacy databases, and other DLT platforms, facilitating the smooth flow of data and transactions across different environments.

- **Programmability**: Corda's smart contract model, called **CorDapps**, provides a flexible and secure framework for defining and enforcing business agreements. CorDapps are mainly written in *Kotlin*, a modern, statically typed programming language that runs on the Java Virtual Machine. Kotlin is designed to be fully interoperable with Java, which means that CorDapps can also be developed using *Java*. This approach enables the automation and execution of complex workflows while maintaining a high level of control and customization.

- **Regulatory compliance**: Corda is designed to meet the stringent regulatory requirements of industries such as finance and healthcare. It incorporates regulatory and legal constructs into the architecture, enabling compliance with data protection, **Know Your Customer (KYC)**, **Anti-Money Laundering (AML)**, and other regulatory frameworks. Corda's focus on privacy and fine-grained access control aligns with regulatory expectations regarding data confidentiality and consent management.

- **Network membership and governance**: Corda networks are permissioned, meaning that participants must be explicitly invited and granted access to the network. This membership model ensures that only trusted entities are part of the network, enhancing security and reducing the risk of malicious activities. Corda also allows for flexible network governance, enabling participants to define and enforce rules and consensus mechanisms, as well as upgrade protocols as per their specific requirements.

Exploring AKS and its benefits

AKS is a managed container orchestration service provided by Microsoft Azure. It simplifies the deployment, management, and scaling of containerized applications using Kubernetes, an open source container orchestration platform. AKS offers a range of benefits that make it an ideal choice for hosting and managing Corda distributed ledger networks.

Overall, AKS offers a powerful and flexible infrastructure for hosting Corda distributed ledger networks. The following figure shows the list of benefits of running a Corda network on AKS, from simplified deployment to scalability, high availability, security features, and monitoring capabilities. Lastly, its integration with Azure services makes AKS an ideal choice for organizations looking to leverage Corda's features while ensuring operational efficiency and reliability:

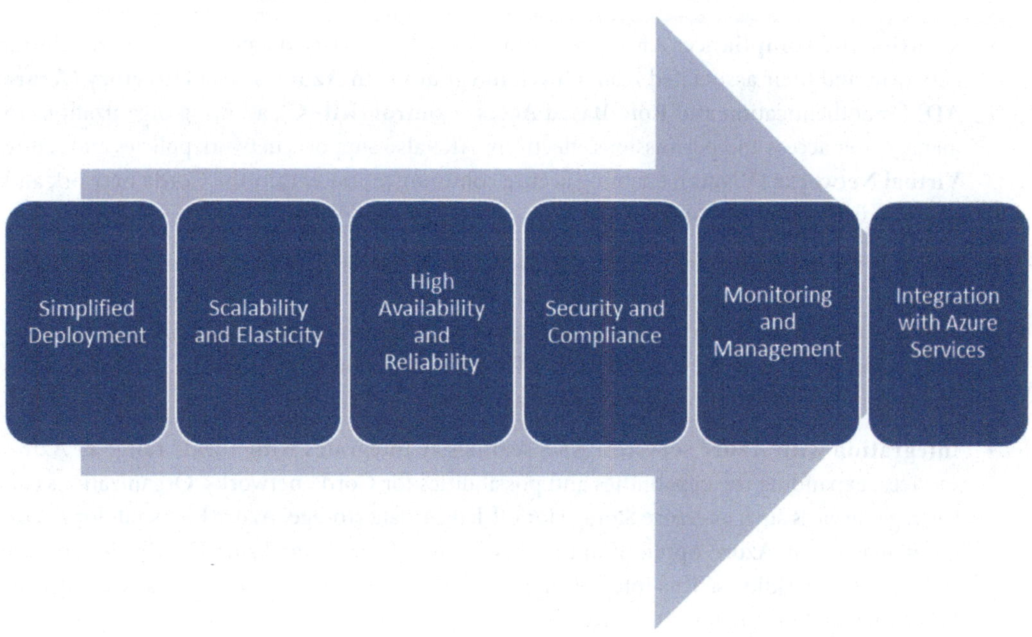

Figure 7.2 – Benefits of AKS

Let's expand on the benefits of AKS with additional details:

- **Simplified deployment**: AKS abstracts away the complexities of setting up and managing Kubernetes clusters. It provides a user-friendly interface and a seamless deployment experience, allowing developers and DevOps teams to focus on their applications rather than infrastructure management. With just a few clicks or commands, AKS provisions and configures a fully functional Kubernetes cluster, eliminating the need for manual setup and configuration.

- **Scalability and elasticity**: AKS enables organizations to effortlessly scale their Corda networks based on demand. It supports horizontal scaling by automatically adding or removing computing resources based on workload requirements. This ensures that Corda nodes can handle increasing transaction volumes and adapt to changing business needs without disruptions. AKS also provides built-in support for load balancing, allowing for efficient distribution of traffic across Corda nodes.

- **High availability and reliability**: AKS ensures the availability and reliability of Corda networks through its fault-tolerant architecture. It automatically manages the placement of Corda nodes across multiple availability zones, minimizing the impact of hardware failures or network disruptions. AKS also offers features such as automatic node repair, rolling updates, and self-healing capabilities, ensuring that the Corda network remains highly available and resilient.

- **Security and compliance**: AKS incorporates robust security measures to protect Corda networks and their associated resources. It integrates with **Azure Active Directory** (**Azure AD**) for authentication and **Role-Based Access Control** (**RBAC**), allowing organizations to manage user access and permissions effectively. AKS also supports network policies and Azure **Virtual Networks** (**VNets**), enabling secure communication within the Corda network and isolation from other resources.

- **Monitoring and management**: AKS provides comprehensive monitoring and management capabilities for Corda networks. It integrates seamlessly with Azure Monitor, allowing organizations to gain insights into the performance, health, and usage of their Corda nodes. AKS also supports integration with Azure DevOps and other DevOps tools, enabling efficient CI/CD workflows and streamlined application life cycle management.

- **Integration with Azure services**: AKS seamlessly integrates with a wide range of Azure services, expanding the capabilities and possibilities for Corda networks. Organizations can leverage services such as Azure Storage for off-ledger data storage, Azure Key Vault for secure key management, Azure Application Insights for monitoring, and Azure DevOps for end-to-end DevOps workflows. This integration provides a comprehensive and cohesive ecosystem for hosting and managing Corda networks.

By hosting Corda on AKS, organizations can leverage the strengths of both platforms to establish a resilient, scalable, and secure infrastructure for Corda-based applications. The synergy between Corda and AKS enables organizations to focus on building and deploying Corda networks while taking advantage of AKS's management capabilities, scalability, and integration with Azure services, ultimately accelerating the development and deployment of Corda-powered solutions.

Architecting Corda networks on AKS

When architecting a Corda network on AKS, several design considerations can contribute to a well-structured and efficient deployment. By considering these design considerations, you can create a well-architected Corda network deployment on AKS that aligns with your business requirements, scalability needs, security measures, and regulatory compliance obligations.

Design considerations for Corda network deployment

The following diagram describes the key design considerations for Corda network deployment on AKS:

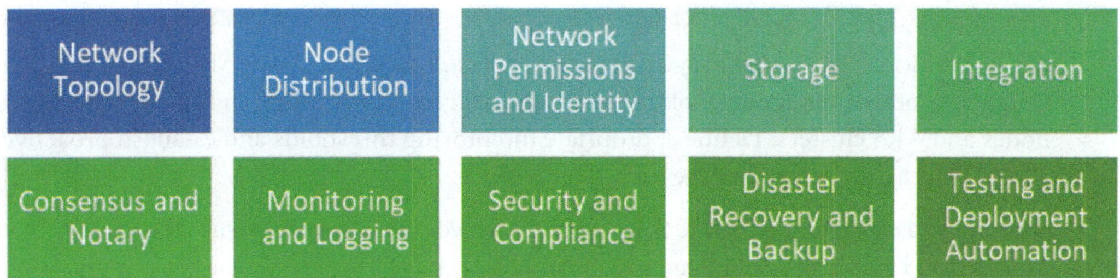

Figure 7.3 – Key design considerations for Corda on AKS

Let's examine each feature in detail:

- **Network topology**: Consider the network topology that best suits your Corda network requirements. Determine whether a single Corda network is sufficient or if multiple networks or subnetworks are necessary to isolate different business domains or participant groups. Define the network boundaries and the required connectivity between Corda nodes within and across networks.

- **Node distribution**: Determine the distribution of Corda nodes across AKS clusters. Consider factors such as fault tolerance, latency, and regulatory requirements. Distributing nodes across multiple availability zones or regions enhances fault tolerance and resilience. Evaluate the trade-offs between centralized versus decentralized node deployments based on network requirements.

- **Network permissions and identity**: Define the membership and permissions model for the Corda network. Identify the participants and their roles, and establish the necessary identity and access controls. Leverage Azure AD or other identity management systems for user authentication and authorization. Utilize Corda's built-in identity framework to manage the cryptographic identities of Corda nodes.

- **Storage**: Determine the storage requirements for Corda network data. Evaluate whether Azure Storage services, such as Azure Blob Storage or Azure Files, are suitable for storing off-ledger data or other data associated with Corda nodes. Consider data encryption and access controls to ensure data privacy and security.

- **Networking and integration**: Plan the networking aspects and integration points of the Corda network with other Azure services or external systems. Determine the connectivity requirements between Corda nodes, client applications, and external services. Define the necessary network security groups, VNet configurations, and firewall rules to ensure secure communication within the network.

- **Consensus and notary configuration**: Evaluate the consensus and notary configuration options based on the Corda network's requirements. Corda allows flexibility in choosing the consensus algorithm and the type of notary service (for example, single, non-validating, or validating). Consider the desired level of decentralization, performance needs, and regulatory compliance when configuring consensus and notary services.

- **Monitoring and logging**: Plan for monitoring and logging capabilities to gain insights into the health, performance, and behavior of the Corda network. Leverage Azure Monitor, Application Insights, or other monitoring tools to collect and analyze metrics, logs, and alerts from Corda nodes and AKS clusters. Define appropriate monitoring thresholds and establish proactive alerting mechanisms for critical events.

- **Security and compliance**: Ensure that the Corda network deployment aligns with security best practices and regulatory compliance requirements. Establish secure network communication channels, enforce encryption at rest and in transit, and implement proper access controls. Consider data privacy regulations, cryptographic key management, and compliance frameworks applicable to your industry.

- **Disaster recovery and backup**: Define disaster recovery and backup strategies to mitigate the impact of unforeseen events. Implement mechanisms to perform regular backups of Corda node data and ensure the availability of backups in case of data loss or system failures. Consider data replication across different regions or cloud providers for enhanced resilience.

- **Testing and deployment automation**: Establish automated testing and deployment processes to streamline the deployment and management of Corda networks on AKS. Utilize CI/CD pipelines, **Infrastructure as Code (IaC)** tools, and testing frameworks to facilitate rapid and consistent deployment, testing, and updating of Corda nodes.

Integrating your AKS configuration for Corda with additional Azure services

Selecting the appropriate AKS configuration for hosting Corda involves considering various factors to ensure optimal performance, scalability, and cost-efficiency. First, you may want to evaluate the resource requirements of your Corda network nodes while considering factors such as transaction volume, computational complexity, and memory requirements. Choose AKS node sizes that provide sufficient CPU and memory resources to support Corda nodes effectively. Additionally, consider the scalability requirements of your Corda network and select AKS node sizes that can accommodate future growth and handle increased transaction volumes.

More details about provisioning an AKS cluster for Corda are described in the following sections.

High availability

When running a service in Azure, you can consider the redundancy of your infrastructure by adopting techniques such as availability sets or availability zones. With either approach, you must first determine the number of Corda nodes required for your network and then distribute them across AKS nodes. As a good practice, consider deploying Corda nodes across multiple availability zones for improved fault tolerance and high availability. Distributing nodes across different availability zones ensures that the network remains operational, even in the event of infrastructure failures or disruptions.

> **Integration with Azure**
>
> Configure load balancing to ensure efficient distribution of traffic across Corda nodes. Utilize Azure Load Balancer or Azure Application Gateway to distribute incoming requests to the Corda network endpoints. Load balancing helps improve the scalability, availability, and fault tolerance of the Corda network. Consider using Azure Private Link to securely access Corda nodes from within your virtual network. Private Link provides private connectivity to your Corda nodes over the Azure backbone network, enhancing security and isolating the network traffic from the public internet.

Follow these steps to configure load balancing in AKS and ensure efficient distribution of traffic across Corda nodes. By setting up a Kubernetes service with the `LoadBalancer` type, AKS will automatically provision an Azure Load Balancer to distribute incoming traffic across your Corda nodes. This load balancing mechanism ensures that your Corda network remains highly available, scales efficiently, and provides optimal performance to clients accessing the various Corda services:

1. After creating a Kubernetes service for your Corda nodes, AKS will abstract the network access to your pods and provide a stable IP address or DNS name for clients to access the Corda nodes. At this point, you can specify the service type for load balancing, which is `LoadBalancer`. This will instruct AKS to automatically provision an Azure Load Balancer and distribute incoming traffic across the Corda nodes.

2. You can set the service type to `LoadBalancer` in AKS using PowerShell. You must use the `Set-AzAksService` cmdlet to update the service type for your Corda network's Kubernetes service to `LoadBalancer`:

```
Set-AzContext -Name <AKSContextName> -SubscriptionId
<SubscriptionId>
Set-AzAksService -ResourceGroupName <ResourceGroupName> -Name
<AKSClusterName> -ServiceName <ServiceName> -Type LoadBalancer
```

   Replace the `<ResourceGroupName>`, `<AKSClusterName>`, and `<ServiceName>` placeholders with the appropriate values for your environment.

3. AKS will create a basic Azure Load Balancer. You can now customize the load balancer's configuration to suit your specific requirements. For example, you can configure session affinity and health probes or use Azure Application Gateway for additional functionality.

   Session affinity (also known as sticky sessions) ensures that requests from a specific client are consistently directed to the same backend pod, improving the user experience when the application requires session state.

4.   Use the `Set-AzAksService` cmdlet to update the service configuration and enable session affinity. The key parameter for enabling session affinity is `SessionAffinity`; it should be set to `ClientIP` to enable sticky sessions based on the client's IP address:

```
Set-AzAksService -ResourceGroupName <ResourceGroupName> -Name
<AKSClusterName> -ServiceName <ServiceName> -SessionAffinity
ClientIP
```

Replace `<ResourceGroupName>`, `<AKSClusterName>`, and `<ServiceName>` with the appropriate values for your environment.

5.   To verify that session affinity has been enabled for the service, you can get the details of the service using the `Get-AzAksService` cmdlet:

```
Get-AzAksService -ResourceGroupName <ResourceGroupName> -Name
<AKSClusterName> -ServiceName <ServiceName>
```

Look for the `SessionAffinity` property. It should be set to `ClientIP`.

Storage

Azure offers multiple options for data storage. When implementing your Corda solution, you may want to assess the storage requirements of your Corda network. Determine if Azure Disk Storage or Azure Files can adequately fulfill the storage needs of Corda nodes. Evaluate factors such as read/write performance, data durability, and backup options. Also, consider whether encryption at rest and data redundancy are necessary based on your security and compliance requirements.

> **Integration with Azure**
>
> Leverage Azure Storage services, such as Azure Blob Storage or Azure Files, for off-ledger data storage or file sharing among Corda nodes. Use Azure Storage SDKs or APIs to interact with the storage services directly from your Corda applications. Securely manage cryptographic keys and secrets using Azure Key Vault. Integrate Corda nodes with Azure Key Vault to securely store and retrieve sensitive information, such as cryptographic keys or passwords, used within the Corda network.

Scalability

Specifically to AKS, the Kubernetes cluster configuration defines the appropriate number of nodes, their size, and the availability zone distribution. Factors such as cluster autoscaling to automatically adjust the number of nodes based on resource utilization and traffic demands are key to delivering a robust DLT solution based on Corda.

> **Integration with Azure**
>
> Leverage Azure DevOps or GitHub services for seamless CI/CD workflows, version control, and release management of Corda applications and network updates. Integrate Corda network deployments with Azure DevOps pipelines for automated testing, building, and deploying Corda nodes and associated applications. Integrate Corda nodes with Azure Event Grid to enable event-driven architectures. Publish events from Corda nodes to Azure Event Grid, which can then trigger downstream processes or invoke Azure Functions for automated responses or further processing. Utilize Azure Logic Apps to automate workflows and orchestrate actions based on specific events or triggers within the Corda network. Integrate Corda nodes with Azure Logic Apps to create custom workflows, automate data processing, or trigger notifications and alerts.

Security and compliance

When speaking of security and compliance, the options in Azure vary and a lot depends on your existing Azure implementation. For example, you may want to enable Azure AD integration for authentication and authorization of AKS resources and Corda network access and leverage AKS features such as RBAC network policies and Azure Security Center integration to enhance your security and compliance posture.

> **Integration with Azure**
>
> Utilize **Network Security Groups** (**NSGs**) and VNet service endpoints to restrict access to Corda nodes. Apply firewall rules and network policies to control inbound and outbound traffic and enforce secure communication within the network. Consider implementing network traffic encryption using **Transport Layer Security** (**TLS**) for enhanced security.

Monitoring and logging

We have already mentioned various options for monitoring and logging in the context of AKS. Gaining insights into the health and performance of Corda nodes and AKS clusters is essential to your solution in the long term. Best practices include leveraging Azure Monitor, Azure Application Insights, or other monitoring tools to collect metrics, logs, and alerts. Also, consider integrating with logging and monitoring solutions specific to Corda, such as Corda Firewall or Corda Monitoring Toolkit, for deeper visibility into Corda network operations.

> **Integration with Azure**
>
> Integrate Corda nodes with Azure Monitor and Azure Application Insights for comprehensive monitoring and diagnostics. Monitor the performance, health, and availability of Corda nodes, capture and analyze logs, and set up alerts and notifications for critical events or anomalies within the Corda network.

Cost optimization

Every resource comes with a cost, so you may also want to look at the best strategy for cost optimization. Apart from wanting to optimize costs by selecting the AKS configuration that best aligns with your budget and resource utilization, you can also do the following:

- Consider reserved instances or spot instances for cost savings

- Evaluate AKS scaling options, including **Horizontal Pod Autoscaling** (**HPA**), to automatically adjust resources based on workload demands

- Monitor and analyze resource utilization to right-size your AKS cluster and nodes effectively

Enabling HPA in AKS requires some additional steps beyond a single PowerShell cmdlet. While you can configure and manage AKS resources, including deployments and ReplicaSets, using PowerShell, enabling HPA is typically done through Kubernetes manifests or the Azure CLI. Let's perform the following steps to enable HPA in AKS:

1. Ensure you have the Azure CLI and Kubernetes CLI (kubectl) installed on your machine. The Azure CLI is used to interact with AKS resources, while kubectl is used to interact with the Kubernetes cluster.

2. Use the `Connect-AzAccount` cmdlet in PowerShell to sign in to your Azure account. Then, use the `az aks get-credentials` command (Azure CLI) or `kubectl config use-context` command (kubectl) to connect to your AKS cluster.

3. HPA relies on **Metrics Server** to collect resource utilization data. If your AKS cluster does not have Metrics Server enabled, you need to install it. Use the `az aks enable-addons --addons monitoring` command to enable the monitoring add-on, which includes Metrics Server.

4. Create a Kubernetes HPA manifest in YAML format. The HPA manifest should specify the deployment or replica set you want to autoscale, as well as the minimum and maximum number of replicas and the metrics used for scaling.

5. Use the `kubectl apply -f` command to apply the HPA manifest to your AKS cluster and enable autoscaling for the specified deployment or replica set.

It is important to note that the specific AKS configuration for Corda may vary based on your network's unique requirements and expected workload. It is recommended to perform thorough testing and performance benchmarking to validate the chosen configuration before deploying Corda networks in production on AKS.

Provisioning an AKS cluster for Corda

The previous two sections focused on understanding the core features of Corda and AKS and the synergy of the two services in Azure, as well as the key considerations for designing a robust, secure, and scalable solution.

We will now progress with the specific steps to set up such a solution based on an AKS cluster running a Corda DLT, starting from having access to an Azure subscription and then preparing the prerequisites for provisioning an AKS cluster for Corda.

Setting up an Azure subscription and preparing the prerequisites

Perform the following steps:

1. Set up an Azure subscription:

   I. Open the Azure portal (`https://portal.azure.com/`) and sign in with your Microsoft account. If you don't have an account yet, you can create a new one for free.

   II. Follow the prompts to create a new Azure subscription or associate an existing subscription with your account.

   III. Provide the necessary information and complete the subscription setup process.

2. Install the Azure CLI.

   Download and install the Azure CLI on your local machine by following the instructions provided in the official Azure CLI documentation at `https://docs.microsoft.com/en-us/cli/azure/install-azure-cli`.

3. Create a resource group:

   I. Open a command prompt or terminal and sign into the Azure CLI by running the following command:

   ```
   az login
   ```

   II. Follow the prompts to authenticate with your Azure account.

   III. Create a resource group to hold the AKS cluster resources by running the following command:

   ```
   az group create --name <resource-group-name> --location
   <location>
   ```

4. Register the required Azure providers.

   Run the following commands to register the required Azure providers:

   ```
   az provider register --namespace Microsoft.ContainerService
   az provider register --namespace Microsoft.Network
   az provider register --namespace Microsoft.Compute
   az provider register --namespace Microsoft.Storage
   ```

5. Create an **Azure Container Registry (ACR)**.

    Run the following command to create an ACR in your resource group:

    ```
    az acr create --resource-group <resource-group-name> --name
    <acr-name> --sku Basic
    ```

6. Create a VNet to isolate your AKS cluster.

    Run the following command to create a VNet in your resource group:

    ```
    az network vnet create --resource-group <resource-group-name>
    --name <vnet-name> --address-prefixes <vnet-address-space>
    --subnet-name <subnet-name> --subnet-prefix <subnet-address-
    prefix>
    ```

7. Create a service principal.

    Generate a service principal for authenticating AKS with Azure resources. Run the following command to create the service principal and assign it the necessary role:

    ```
    az ad sp create-for-rbac --skip-assignment --name <service-
    principal-name>
    ```

    Take note of the `appId` (client ID) and `password` (client secret) values returned by the preceding command.

8. Grant permissions to the service principal.

    Assign the necessary permissions to the service principal by running the following commands:

    ```
    az role assignment create --assignee <service-principal-appId>
    --role "Contributor" --scope /subscriptions/<subscription-id>/
    resourceGroups/<resource-group-name>

    az role assignment create --assignee <service-principal-appId>
    --role "Contributor" --scope /subscriptions/<subscription-id>/
    resourceGroups/<resource-group-name>/providers/Microsoft.
    ContainerRegistry/registries/<acr-name>
    ```

9. Create a Kubernetes secret to store the service principal credentials securely.

    Run the following command to create the secret in your AKS cluster:

    ```
    kubectl create secret generic corda-azure-credentials \
            --from-literal=subscriptionId=<subscription-id> \
            --from-literal=tenantId=<tenant-id> \
            --from-literal=clientId=<service-principal-appId> \
            --from-literal=clientSecret=<service-principal-password>
    ```

    Replace `<subscription-id>`, `<tenant-id>`, `<service-principal-appId>`, and `<service-principal-password>` with the respective values.

10. Prepare the Corda configuration files.

    Gather the necessary Corda configuration files, such as network parameters and node configuration files, according to your Corda network requirements.

Once you have completed these steps, you will have set up an Azure subscription, prepared the prerequisites, and created the necessary resources to provision an AKS cluster for Corda. You can now proceed with creating the AKS cluster using the Azure portal, Azure CLI, or other deployment methods of your choice.

Deploying an AKS cluster with the appropriate Corda configuration

Once the initial setup of Azure resources and necessary prerequisites is complete, you can now deploy an AKS cluster with the appropriate Corda configurations:

1. Define the Corda node configuration.

   Prepare the Corda node configuration files (for example, `node.conf`) for each Corda node in your network. Customize the configuration according to your network's requirements, including network parameters, notary settings, identity information, and other relevant parameters.

2. Create a Kubernetes deployment YAML:

   I.   Create a Kubernetes deployment YAML file (`corda-deployment.yaml`) to define the deployment of Corda nodes in the AKS cluster.

   II. Specify the container image for Corda nodes, as well as the desired number of replicas, resource limits, and environment variables for Corda node configurations.

   III. Mount the necessary configuration files (for example, `node.conf`) as ConfigMaps or secrets to provide them to the Corda nodes.

3. Apply the Kubernetes deployment.

   Use the `kubectl apply` command to deploy the Corda nodes in the AKS cluster by applying the deployment YAML file:

   ```
   kubectl apply -f corda-deployment.yaml
   ```

4. Monitor the deployment progress using `kubectl` commands, such as `kubectl get pods` or `kubectl get deployment`, to ensure that the Corda nodes are successfully created and running.

5. Access the Corda nodes:

   I.   Obtain the external IP address or DNS name of the Corda nodes by running `kubectl get service` or going through the Azure portal.

   II. Use the obtained IP address or DNS name to connect to the Corda nodes from external clients or other network participants.

6. Validate Corda network connectivity:

    I.   Validate the connectivity between Corda nodes by initiating transactions, sending messages, or performing other interactions within the Corda network.

    II. Monitor logs, metrics, and Corda-specific tools to ensure the proper functioning and communication of the Corda nodes.

7. Scale and manage the AKS Cluster:

    I.   Use AKS's scaling capabilities to scale the Corda nodes based on your network's requirements and load. You can scale horizontally by adjusting the replica count in the deployment YAML or using AKS's autoscaling feature.

    II. Employ AKS management capabilities to monitor, upgrade, and manage the AKS cluster, including node scaling, rolling updates, and version upgrades.

8. Implement backup and disaster recovery:

    I.   Set up regular backups of Corda node data and relevant configuration files using appropriate backup mechanisms. Consider using Azure Backup or other backup solutions.

    II. Implement disaster recovery strategies, such as replicating data across different regions or using Azure Site Recovery, to ensure business continuity in case of unexpected failures or outages.

    III. Throughout the deployment process, closely monitor the health, performance, and connectivity of the Corda network. Regularly review logs, metrics, and monitoring tools to identify and address any issues that may arise.

It's important to note that the specific deployment process may vary, depending on your specific Corda network requirements and the tools and methodologies you choose to use. Always refer to the Corda and Kubernetes documentation for detailed instructions and best practices related to the deployment and management of Corda nodes on AKS.

Corda Enterprise in the Azure Marketplace

The approach described in the previous section implies that you are comfortable managing the AKS cluster where the Corda network will be deployed. That is certainly the deployment method with the most control over the provisioned infrastructure.

However, especially for testing or evaluation purposes, there may be circumstances in which you may want a quicker way to deploy a Corda network without the hassle of managing the underlying infrastructure.

If you want to accelerate the deployment of a Corda network, R3 offers a predefined template in the Azure Marketplace for provisioning a Corda Enterprise single node for Corda Testnet. This template deploys a single Corda Enterprise node and helps the user obtain a certificate for Corda Testnet, a live network of global nodes that can be transacted for testing purposes. The following screenshot shows the **Corda Enterprise on Corda Testnet** service in the Azure Marketplace:

Figure 7.4 – The Corda Enterprise service in the Azure Marketplace

As reported by R3 on the Azure Marketplace, by using this service, you can quickly deploy Corda Enterprise nodes and test applications on a network of Corda nodes that have access to Corda Testnet. Here are some of the key features of this deployment of Corda Enterprise:

- **Corporate-level security**: Corda Enterprise nodes are deployed inside corporate data centers and still retain the ability to communicate securely with other nodes.

- **Off-chain data storage**: Seamless integration with database services, such as Azure SQL and Oracle database.

- **Runs on Corda**: You can operate cross-version networks of Corda nodes by seamlessly integrating corporate firewall-protected nodes that support multiple CorDapps on the same network. They are enabled by Corda's privacy model.

- **The benefits of PaaS**: All the benefits that you'd expect from a **Platform as a Service** (**PaaS**) deployment in Azure, including predictable release schedules, performance and availability monitoring, enhanced security, high availability, and disaster recovery of Corda nodes.

Securing the AKS cluster and managing access controls

You're nearly there. The last step in provisioning an AKS cluster for Corda is securing the cluster and managing access controls. These are essential steps to protect your Corda network and ensure proper authorization. The following diagram, sourced from StackSimplify, depicts the connections between Azure AD and the AKS cluster. For additional information on this process, please consult `https://stacksimplify.com/azure-aks/azure-ad-authentication-for-aks-admins/`:

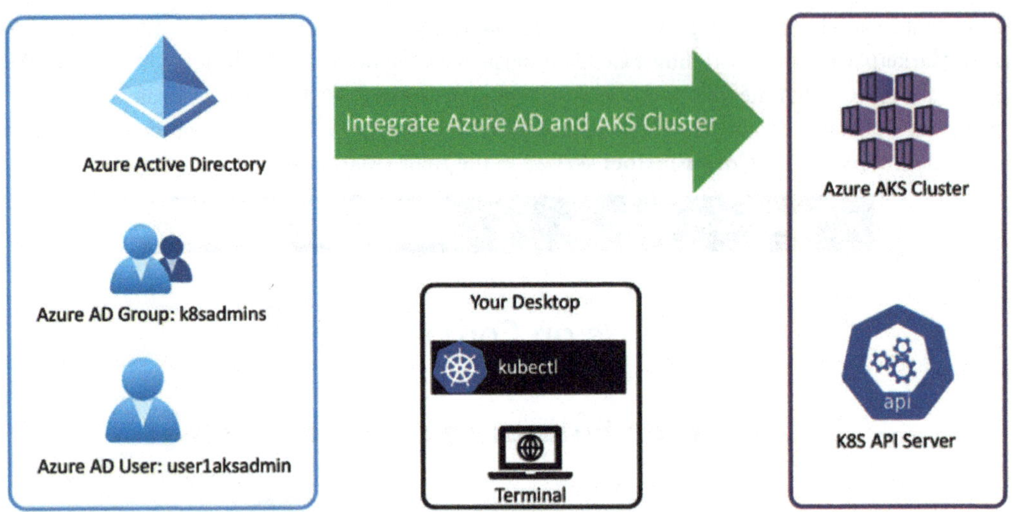

Figure 7.5 – Azure AD authentication for AKS administrators

Step by step, this is how you can configure robust user authentication and role-based authorization in AKS. Follow these steps to ensure you have strong security in your AKS deployment:

1. Enable Azure AD integration:

    I.   Integrate your AKS cluster with Azure AD for authentication and authorization.

    II.   Use the Azure CLI or Azure portal to enable Azure AD integration for your AKS cluster.

2. RBAC:

    I.   Utilize RBAC to manage access controls within the AKS cluster.

    II.   Define custom Azure AD roles or use built-in roles to grant appropriate permissions to users or groups.

    III. Assign roles to users or groups for managing AKS resources, such as cluster administration or pod access.

3. Set up the specific Kubernetes RBAC:

    I.   Leverage Kubernetes RBAC to control access to cluster resources.

    II.   Define Kubernetes Roles and RoleBindings or ClusterRoles and ClusterRoleBindings to grant specific permissions to users or groups within the cluster.

4. Implement network policies:

    I.   Implement network policies to control inbound and outbound traffic within the AKS cluster.

    II.   Use Kubernetes Network Policies to define rules for pod-to-pod communication, allowing only authorized traffic between Corda nodes or specific network segments.

5. Leverage Azure Firewall or NSGs:

    I.   Use Azure Firewall or NSGs to enforce network-level security controls for inbound and outbound traffic to the AKS cluster.

    II.   Define appropriate firewall rules or NSG rules to allow only necessary traffic and block unauthorized access.

6. Enforce TLS:

    I.   Enable TLS for secure communication within the Corda network and between Corda nodes.

    II.   Generate and manage TLS certificates for Corda nodes to ensure encrypted communication.

7. Secrets management:

    I.   Store sensitive information, such as credentials or private keys, securely using Kubernetes Secrets or Azure Key Vault.

    II.   Use secrets to provide secure access to sensitive information required by Corda nodes or other applications running within the cluster.

8. Container image security:

    I.   Employ container image security best practices.

    II.   Use container image scanning tools to identify vulnerabilities and ensure images are free from known security issues before deployment.

9. Regular updates and patching:

    I.   Stay updated with the latest AKS versions and patches.

    II.   Regularly apply updates to the AKS cluster, including security patches, bug fixes, and feature enhancements.

10. Monitoring and logging:

   I.   Implement monitoring and logging solutions to gain visibility into cluster activities and potential security threats.

   II.   Leverage Azure Monitor, Azure Security Center, or other monitoring tools to collect and analyze logs, metrics, and security events.

11. Regular security audits and reviews:

   I.   Conduct periodic security audits and reviews to identify potential vulnerabilities or misconfigurations.

   II.   Regularly assess and update security controls based on the evolving threat landscape and industry best practices.

By following these steps, you can enhance the security of your AKS cluster and effectively manage access controls for your Corda network, ensuring that only authorized users and entities have the necessary permissions to interact with the cluster and Corda nodes.

Managing Corda nodes on AKS

Following the initial provisioning and setup of a Corda network on AKS, your deployed infrastructure now requires constant attention: performance, scalability, high availability, security, and operating costs are just a few of the conditions to regularly check for a sound and robust environment.

Now, let's analyze how to scale, validate the resilience, and ensure high availability and disaster recovery of Corda networks on AKS.

Scaling Corda nodes as Kubernetes pods

Your AKS cluster is now up and running. We'll keep on leveraging the `corda-deployment.yaml` file we created in the previous section when we deployed Corda on Kubernetes pods. This file specifies the desired number of replicas, resource limits, and other configuration options for the Corda nodes. In addition, the YAML file describes the source for the container image for Corda nodes. This can be an image pulled from a container registry or a custom-built image that includes the necessary Corda software and dependencies.

A typical strategy for anticipating traffic demand on your AKS cluster is to create a capacity plan based on the expected growth and workload of the Corda network. Such a plan takes into consideration factors such as transaction volume, number of participants, data growth, and expected network activity over time (first week, first month, 3 months, 6 months, 1 year). This capacity plan with help you estimate the required resources (CPU, memory, and storage) and scalability needs to accommodate future demand.

You can scale Corda nodes as Kubernetes pods on AKS by following these steps:

1. Configure Corda node identity.

   There is no single out-of-the-box script provided by Corda or AKS that specifically generates and configures Corda node identity information for scaling Corda nodes as Kubernetes pods on AKS. The process of generating and configuring Corda node identity information involves several steps and considerations, and it often depends on your specific deployment requirements and security considerations:

   I.   Generate and configure the Corda node identity information (for example, private keys and certificates) as part of the Corda configuration files or secrets. Corda nodes require cryptographic identities to communicate securely within the network. Use Corda's built-in identity generation tools to generate the required cryptographic key pairs and certificates for each Corda node. These tools include `corda.jar` and `network-bootstrapper`.

   II. Ensure that the Corda nodes have the necessary identity information to participate in the Corda network. For example, you may want to use a **Certificate Authority (CA)** to issue and manage the certificates for Corda nodes. A CA helps ensure the authenticity of the node identities and can simplify the certificate management process.

   III. Securely store credentials, such as cryptographic key pairs and certificates. Avoid exposing sensitive information in version control or public repositories. Rather, prefer using **Kubernetes secrets** to store and manage the Corda node identity information securely. AKS secrets enable you to store sensitive data such as cryptographic keys and certificates as Kubernetes objects.

2. Validate Corda network connectivity:

   I.   Validate the connectivity between the Corda nodes running as Kubernetes pods by initiating transactions, sending messages, or performing other interactions within the Corda network. Use **Corda Shell** (`corda-shell`) to interact with your Corda nodes. Corda Shell allows you to issue commands, initiate transactions, and query the network.

   II. Monitor logs, metrics, and Corda-specific tools to ensure the proper functioning and communication of the Corda nodes.

3. Scale and manage the Corda pods:

   I.   Use Kubernetes scaling capabilities to scale the Corda pods based on your network's requirements and load. You can scale horizontally by adjusting the replica count in the deployment YAML or using the Kubernetes autoscaling feature.

   II. Employ Kubernetes management capabilities to monitor, upgrade, and manage the Corda pods, including rolling updates, version upgrades, and pod life cycle management

You can configure your AKS cluster to scale manually or automatically. The two approaches to the scalability of an AKS cluster are described in the following table:

Manual scaling	Auto scaling
<ul><li>Monitor the workload and performance of the Corda network regularly</li><li>Manually adjust the replica count of Corda pods based on the observed load or anticipated changes in network requirements</li><li>Use the `kubectl scale` command to scale the Corda deployment up or down: `kubectl scale deployment corda-deployment --replicas=<replica_count>`</li></ul>	<ul><li>Configure HPA for the Corda nodes within the AKS cluster</li><li>Set thresholds based on metrics such as CPU utilization or pending requests to automatically scale up or down the number of Corda pods</li><li>Define minimum and maximum replica counts to ensure the cluster scales within the desired bounds</li></ul>

Table 7.1 – Manual versus automatic scaling of Corda nodes on AKS

By following these steps, you can configure and scale Corda nodes as Kubernetes pods on AKS. This allows you to leverage the scalability, manageability, and container orchestration features of Kubernetes while running Corda in a distributed and resilient manner.

However, please keep in mind that there is no *magic formula* for guaranteeing the scalability of your solution. In the end, it all comes down to continuous monitoring, testing, and optimization of the deployed resources. Don't forget to continuously analyze performance metrics and identify any bottlenecks or performance issues within the Corda nodes. This will help you optimize the Corda configuration, such as tuning thread pool sizes, database settings, and network parameters, to improve performance.

Load testing is also an essential step in the healthy life cycle of your Corda network. Conduct regular load testing on the Corda network to simulate real-world usage scenarios and evaluate the performance and scalability of the system. You can use tools such as JMeter, Gatling, or custom scripts to generate realistic workloads and measure the response times and throughput of the Corda nodes. Analyzing the load testing results will help you identify performance bottlenecks and determine the optimal scaling requirements.

Testing and validating the resilience of Corda networks on AKS

Testing and validating the resilience of Corda networks on AKS is essential to ensure that your network can withstand various failure scenarios and continue to operate effectively. Make resilience testing a regular practice in your development and maintenance life cycle. Regularly test and validate the resilience of your Corda network as you make updates, introduce new components, or change the configuration.

The nine-step recommended strategy for testing and validating the resilience of Corda networks on AKS, as illustrated in the following figure, is essential for identifying potential weaknesses, optimizing the network's ability to recover from failures, and ensuring that your Corda network can continue to operate reliably in challenging conditions:

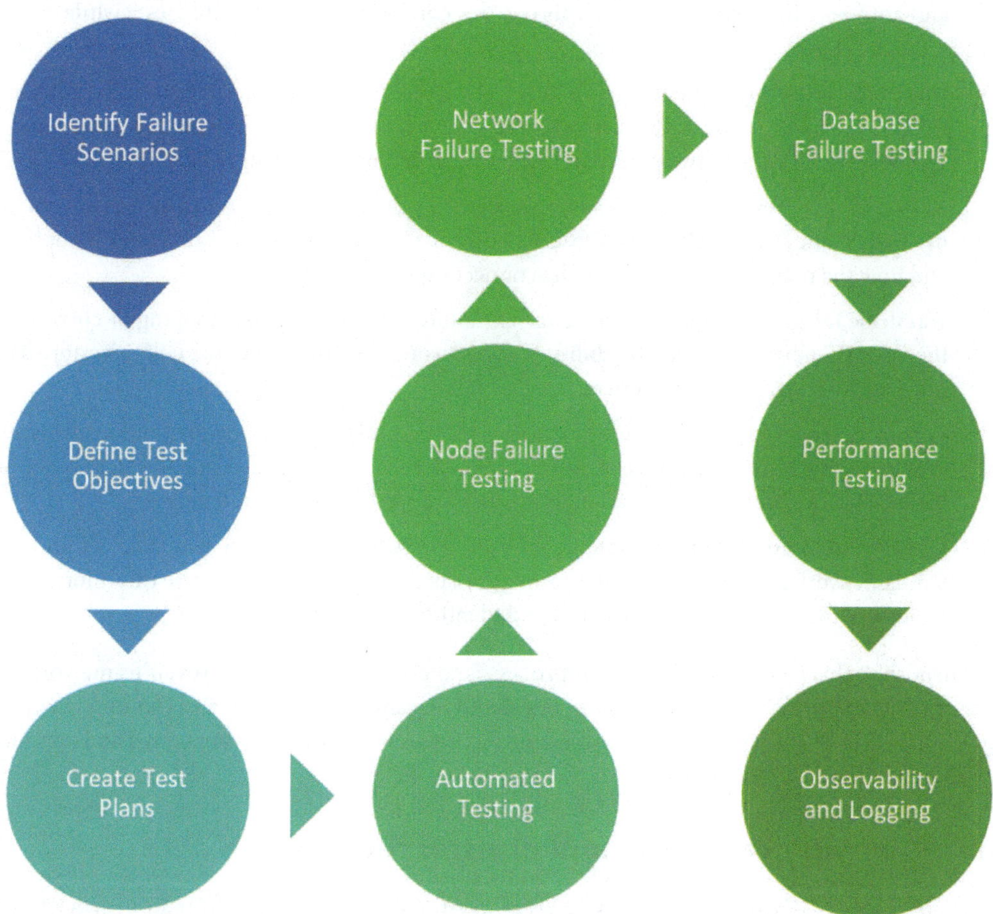

Figure 7.6 – Strategy for testing and validating Corda networks

The strategy that you want to follow for testing and validating Corda networks is as follows:

1. **Identify failure scenarios**: Start by identifying potential failure scenarios that could impact your Corda network. This may include node failures, network failures, database failures, or other critical components.

2. **Define test objectives**: Clearly define the objectives of your resilience testing. Determine what aspects of your Corda network's resilience you want to validate, such as node recovery, data integrity, failover time, or performance under stress.

3. **Create test plans**: Develop test plans that outline specific test cases for each identified failure scenario. Define the steps to simulate the failure, the expected behavior or recovery process, and the metrics or observations to capture during testing.

4. **Automated testing**: Leverage automation tools and frameworks to simulate failure scenarios and automate the testing process. This ensures consistent and repeatable tests while saving time and effort.

5. **Node failure testing**: Simulate node failures by intentionally stopping or terminating individual Corda nodes within the AKS cluster. Monitor the behavior of the network during the failure and observe how the network recovers when the failed node is brought back online.

6. **Network failure testing**: Introduce network failures by disconnecting network links or simulating network latency and packet loss. Observe how the Corda nodes handle network disruptions and how the network recovers once the connection is restored.

7. **Database failure testing**: Simulate database failures by intentionally stopping or corrupting the Corda database. Validate the ability of the network to recover from the failure, restore data integrity, and resume normal operation.

8. **Performance testing**: Conduct performance testing under various load conditions to assess the scalability and performance of your Corda network. Measure response times, transaction throughput, and resource utilization to identify any performance bottlenecks.

9. **Observability and logging**: Implement observability and logging mechanisms within your Corda network to capture detailed information during resilience testing. Use this information to analyze the behavior of the network and identify areas for improvement.

The purpose of this testing and validation process is to provide you with a proven framework for analyzing the results of your resilience testing and identifying any weaknesses or areas for improvement. The action is then to address any issues or bottlenecks that were identified during testing and optimize your Corda network accordingly.

Ensuring high availability and disaster recovery

To ensure high availability and disaster recovery for your Corda deployment on AKS, it's important to implement load balancing and fault tolerance strategies. The following diagram lists a recommended approach:

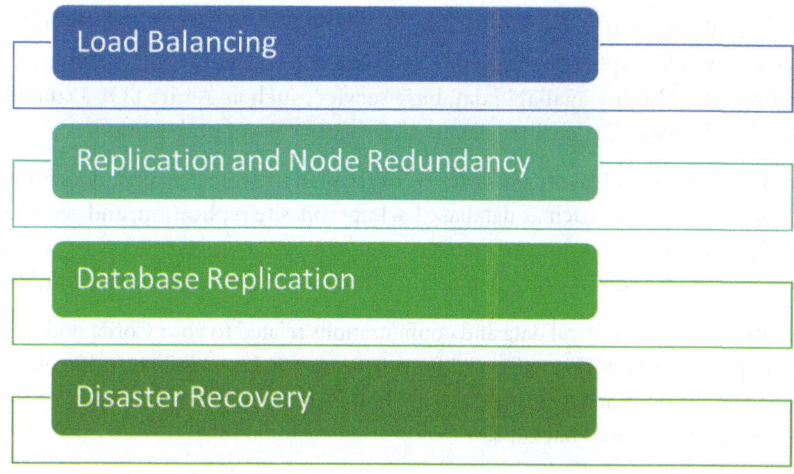

Figure 7.7 – Core characteristics of a highly available cloud application

Let's expand each of these characteristics and map them to the relevant service in Azure:

- **Load balancing**: Implement load balancing to distribute the incoming traffic across multiple Corda nodes. This ensures even distribution of requests and prevents any single node from becoming a performance bottleneck. There are several options for load balancing in AKS:

  - **Azure Load Balancer**: Use Azure Load Balancer, a native Azure service, to distribute incoming traffic across Corda nodes. Azure Load Balancer supports various load balancing algorithms, such as round-robin or least connections, to evenly distribute traffic.

  - **Application Gateway**: Consider using Azure Application Gateway, which operates at the application layer (Layer 7) and provides additional functionalities such as SSL termination, URL-based routing, and session affinity.

  - **Ingress Controller**: Utilize an Ingress controller, such as NGINX Ingress Controller or Azure Application Gateway Ingress Controller, to manage external access and load balancing for your Corda nodes. Ingress controllers allow you to define routing rules and SSL termination for incoming traffic.

- **Replication and node redundancy**: Ensure that you have multiple replica instances of Corda nodes running in your AKS cluster. This provides redundancy and fault tolerance in case any individual node becomes unavailable. Replication can be achieved through Kubernetes deployments and ReplicaSets, allowing automatic scaling and fault recovery.

  In addition, it is recommended to implement health checks and monitoring for the Corda nodes to identify and handle any unhealthy or non-responsive nodes. Use Kubernetes readiness and liveness probes to regularly check the health of the Corda nodes. If a node fails the health check, the load balancer can automatically redirect traffic to healthy nodes.

- **Database replication**: If you're using a separate database for Corda, ensure that the database has been configured for replication and high availability. Replicating the database across multiple nodes or using a highly available database service, such as Azure SQL Database or Azure Database for PostgreSQL, ensures data availability in case of node failures.

- **Disaster recovery**: Implement a disaster recovery plan to protect against major outages or data loss. Consider strategies such as database backups, off-site replication, and geo-redundancy to ensure data durability and disaster recovery capabilities. Azure provides services such as Azure Site Recovery and Azure Backup that can assist in creating robust disaster recovery solutions.

  For regular backups of critical data and configurations related to your Corda nodes (this includes backing up the Corda node configuration, cryptographic keys, and any other essential data), a very good common practice is using Azure Blob storage, Azure File Storage, or Azure Backup for secure and automated backups.

Lastly, do not forget to regularly test your high availability and disaster recovery strategies to ensure they function as expected. Conduct failover tests, simulate node failures, and verify that the load balancing and fault tolerance mechanisms are operating correctly.

Performing failover tests in AKS

Performing failover tests in an AKS cluster involves simulating various failure scenarios and validating how the cluster responds to those failures. It helps ensure the high availability and resilience of your AKS cluster. This section will outline what you can do using PowerShell and the Azure CLI to perform some common failover tests.

> **Before you continue**
>
> Please note that failover tests can affect the availability of your AKS cluster and applications, so it's essential to perform these tests in a controlled environment, such as a staging or test AKS cluster.

Here's a general guide to performing failover tests on an AKS cluster using PowerShell and the Azure CLI:

1. Select the AKS cluster and set the right context.

    Use the `Get-AzAksCluster` cmdlet to list the available AKS clusters and select the one you want to perform the failover tests on. Then, set the AKS context so that it works with that cluster:

    ```
    $aksCluster = Get-AzAksCluster -ResourceGroupName
    <ResourceGroupName> -Name <AKSClusterName>
    Set-AzContext -SubscriptionId $aksCluster.ManagedBy
    Set-AzContext -Name $aksCluster.Name -ResourceGroupName
    $aksCluster.ResourceGroupName
    ```

2. Simulate node failure.

    To simulate node failure, you can scale down a node pool to zero replicas using the Azure CLI. This will drain and delete the nodes in the specified node pool. Observe how the cluster responds and auto-scales to maintain the desired replica count:

    ```
    az aks nodepool scale --resource-group <ResourceGroupName>
    --cluster-name <AKSClusterName> --name <NodePoolName> --node-
    count 0
    ```

3. Test pod rescheduling.

    To test how pods reschedule on node failure, scale the node pool back up to its original replica count using the Azure CLI. Observe how the pods are rescheduled onto the newly added nodes:

    ```
    az aks nodepool scale --resource-group <ResourceGroupName>
    --cluster-name <AKSClusterName> --name <NodePoolName> --node-
    count <OriginalNodeCount>
    ```

4. Simulate cluster availability zone failure (if applicable).

    If your AKS cluster spans multiple availability zones, you can simulate an availability zone failure by draining and deleting all the nodes in one of the zones. Ensure that your node pools are spread across availability zones for this test.

5. Observe cluster recovery.

    After simulating the availability zone failure, the AKS cluster should recover automatically by rescheduling the affected pods onto healthy nodes in the other availability zones.

Please ensure that you fully understand the impact of the failover tests and have proper backup and disaster recovery mechanisms in place before performing them. These steps provided a general guide to help you get started, though you may need to tailor the tests based on your specific AKS cluster configuration and requirements. Always perform these tests in a controlled environment to avoid affecting production workloads.

Summary

Hosting a Corda DLT network on AKS provides organizations with a powerful combination of cutting-edge DLT technology and scalable infrastructure. This chapter served as a comprehensive guide to help you navigate the intricacies of deploying and managing Corda networks on AKS successfully. By following the best practices and guidelines presented here, you can unlock the full potential of Corda and harness the benefits of AKS at scale.

In the next chapter, we'll progress our deep dive into blockchain-related services offered by Azure by looking at using the ledger features of SQL databases. Besides a transactional and relational database, Azure SQL Managed Instance and SQL Server also offer immutable storage facilities with digital ledger technology.

Further reading

Here are some websites where you can learn more about Corda and AKS.

For Corda:

- **Corda Documentation**: The official documentation for Corda provides comprehensive information on getting started, architecture, development, and more. You can find it at `https://docs.corda.net/`.

- **Corda Network**: Visit the Corda Network website to learn more about Corda's global blockchain network, membership, and how to join: `https://www.corda.network/`.

- **Corda GitHub repository**: The Corda GitHub repository contains the open source code for Corda. You can explore the source code, contribute to the project, and access sample applications: `https://github.com/corda/corda`.

- **Corda YouTube channel**: The official Corda YouTube channel features tutorials, webinars, and presentations on various aspects of Corda: `https://www.youtube.com/@Cordablockchain`.

For AKS:

- **AKS documentation**: The official documentation for AKS provides guides, tutorials, and best practices for working with AKS: `https://docs.microsoft.com/azure/aks/`

- **AKS GitHub repository**: The official GitHub repository for AKS contains examples, sample code, and community-contributed content: `https://github.com/Azure/AKS`

- **Microsoft Learn – Kubernetes learning path**: Microsoft Learn offers a Kubernetes learning path that covers AKS and Kubernetes fundamentals: `https://learn.microsoft.com/en-us/azure/aks/concepts-clusters-workloads`

- **Kubernetes documentation**: Familiarize yourself with Kubernetes' concepts and features to better understand AKS: `https://kubernetes.io/docs/`

8

Using the Ledger Features of Azure SQL

In today's data-driven world, maintaining the integrity and immutability of sensitive information is paramount for businesses and organizations. Blockchain technology, with its decentralized ledger system, has set new standards for data security and transparency. While blockchain has found its place in various industries, its implementation can be complex and resource-intensive. To address this, Microsoft Azure offers a more accessible and efficient alternative through the ledger features of Azure SQL Database.

The ledger features of Azure SQL Database bring the advantages of blockchain technology to a broader audience, simplifying data integrity, security, and auditing for businesses of all sizes. With a tamper-evident and immutable transaction log, Azure SQL Database provides a trustworthy foundation for applications that require an extra level of assurance and transparency. Whether it's financial systems, supply chain management, healthcare, or any other industry requiring robust data protection, Azure SQL Database's Ledger offer a powerful and accessible solution to safeguard critical information.

Throughout this chapter, we'll familiarize ourselves with the core concepts of ledgers and blockchain technology since the ledger features in Azure SQL Database are designed to mimic some of these principles, such as immutability and cryptographic hashing.

We'll also plan our data model so that it accommodates the use of ledger features. While Azure SQL Ledger maintains a transaction log for you, you need to design your database schema in a way that aligns with your data integrity and auditing requirements.

We must also be aware of the potential impacts on storage and performance when using ledger features as they involve maintaining a history of transactions.

Lastly, we will understand our data retention policies and regulatory compliance requirements since the use of ledger features may have implications on data archival and compliance audits.

Let's dive into the following main topics:

* Introduction to the ledger features of Azure SQL

* Benefits of the ledger features of Azure SQL

* Using the ledger features of Azure SQL for blockchain solutions

* Integrating the ledger features of Azure SQL with other Azure services

* Best practices for implementing blockchain solutions with the ledger features of Azure SQL

Technical requirements

Before you start using the ledger features in Azure SQL Database, there are some technical prerequisites that you should meet to ensure the smooth implementation and utilization of these features. First, you need an active Microsoft Azure subscription to access and use Azure SQL Database and its features. If you don't have one, you can sign up for a free trial or a pay-as-you-go subscription at `https://azure.com`.

Once you've obtained an Azure subscription, you need an Azure SQL Database instance provisioned and running. The ledger feature is available for SQL Server 2022, Azure SQL Database, and Azure SQL Managed Instance. Specifically, the ledger features are available in the Business Critical and Premium service tiers of Azure SQL Database.

Finally, make sure that you have appropriate permissions to manage Azure SQL Database and configure its features. Typically, this requires the role of a Database Administrator or being the owner of the Azure SQL Database resource.

Introduction to the ledger features of Azure SQL

Azure SQL Database, a fully managed relational database service by Microsoft, provides a scalable, secure, and high-performance platform for applications running on Microsoft Azure. With its continuous evolution, Microsoft introduced ledger capabilities to Azure SQL Database, empowering businesses to benefit from the robustness of blockchain technology without the complexities typically associated with blockchain deployment.

At its core, the ledger features of Azure SQL Database enable organizations to create immutable and tamper-evident records of their data transactions. It ensures data integrity by providing a secure and auditable log of all changes made to the database. Unlike traditional databases, where data changes can be overwritten or deleted without a trace, the ledger features maintain a transparent history of each alteration.

This approach not only enhances data security but also fosters trust among stakeholders. The cryptographic hashing of transactions creates an ordered sequence of blocks of transactions, ensuring that no single entity can manipulate the data retroactively. This feature is particularly valuable in scenarios where regulatory compliance and data auditing are critical requirements.

> **Azure SQL Ledger is not a blockchain**
>
> While the concept of a sequence of blocks may sound similar to a blockchain network, it's essential to clarify that Azure SQL Ledger is not a decentralized and distributed digital ledger with a consensus mechanism.
>
> Azure SQL Ledger and blockchain are distinct technologies that are designed to ensure data integrity and trust, but they operate on different principles and serve varied use cases. Azure SQL Ledger provides cryptographic verifiability of database changes, allowing users to maintain an immutable record of data modifications within a centralized SQL database environment. It offers a ledger layer on top of a relational database.
>
> On the other hand, blockchain is a decentralized ledger technology that distributes data across a network of computers, ensuring transparency, security, and integrity through consensus mechanisms among participants. While Azure SQL Ledger emphasizes enhancing traditional database systems with verifiability features, blockchain focuses on decentralization and peer-to-peer interaction, making it more suited for scenarios that require distributed trust and elimination of central authority.

Let's summarize the key benefits of using the ledger features of Azure SQL in blockchain applications:

- **Security and integrity**: The ledger features of Azure SQL Database offer an additional layer of security by design. The data remains protected against unauthorized modifications, making it suitable for applications that require strict compliance and data governance.

- **Simplified auditing**: The ledger provides a comprehensive transaction history, simplifying the auditing process. Organizations can easily trace back to specific points in time and investigate the sequence of events leading to any particular data state.

- **Data immutability**: Once a transaction has been committed to the ledger, it becomes immutable. This feature eliminates the risk of accidental or malicious data alteration and enhances the credibility of records.

- **Performance and scalability**: Azure SQL Database's ledger features are built to deliver high performance and scalability. Businesses can handle large volumes of data transactions efficiently without compromising on speed and responsiveness.

- **Integration with existing applications**: The ledger features are designed to seamlessly integrate with existing applications. This means organizations can leverage the benefits of blockchain-like data integrity without significant changes to their current infrastructure.

These ledger features are available in certain editions of Azure SQL Database. Specifically, they are available in the Business Critical and Premium service tiers. Be sure to check out the latest Azure SQL Database documentation for any updates on supported editions: `https://learn.microsoft.com/en-us/azure/azure-sql/database/ledger-landing`.

It's important to remember that, while the ledger features abstract many complexities, you should have a good understanding of **Transact-SQL (T-SQL)**, the SQL language variant used with Azure SQL Database. This will help you interact with the database and utilize the Ledger features effectively.

Use cases for SQL Ledger

Ledger features in Azure SQL Database are well-suited for various use cases where data immutability, transparency, and audibility are critical requirements. Typical use cases for ledger technology span across multiple industries, such as the following:

- **Financial transactions**: Ledger technology is widely used in financial systems to record and track transactions, ensuring that financial data remains tamper-proof and auditable. It helps maintain an accurate and transparent ledger of financial activities, which is crucial for regulatory compliance and fraud prevention.

- **Supply chain management**: In supply chain management, ledgers can be employed to track the movement of goods and raw materials from their source to the end user. The immutable records provided by the ledger enhance trust and transparency between parties and help identify inefficiencies or bottlenecks in the supply chain.

- **Healthcare records**: In the healthcare industry, ledgers can be utilized to securely manage patient records, medical history, and treatment information. The immutable nature of the ledger ensures data integrity, preventing unauthorized changes to patient records and supporting data sharing among authorized healthcare providers.

- **Intellectual property rights**: Ledgers can be used to manage intellectual property rights, such as patents, copyrights, and trademarks. The ledger maintains a transparent history of ownership and licensing, providing an auditable trail for legal and licensing purposes.

- **Identity and access management**: Ledger technology can be applied to manage identity and access control. Immutable records of user access and permissions help ensure data security and prevent unauthorized access.

- **Public sector governance**: In the public sector, ledgers can enhance transparency and accountability in government operations. They can be used to track public funds, expenditures, and contract management, reducing the potential for corruption and enhancing citizen trust.

- **Legal and compliance records**: Ledgers provide an ideal solution for legal and compliance records, such as contracts, agreements, and regulatory filings. The transparent and tamper-evident nature of the ledger ensures the integrity and validity of critical legal documents.

- **Real estate and property transactions**: In real estate, ledgers can be employed to record property transactions, ownership history, and title transfers. This simplifies the process of verifying property ownership and reduces the risk of fraudulent property transactions.

- **Supply chain traceability**: Ledgers can facilitate supply chain traceability for products, especially in industries where consumers demand transparency about the origin and production process of the goods they purchase.

- **Voting systems**: Ledger technology can be used to create secure and tamper-proof voting systems, ensuring the integrity of election results and providing transparent audit trails.

These are just a few examples of the diverse applications of ledger technology. The underlying principles of data immutability, transparency, and cryptographic security make ledgers an attractive solution for a wide range of industries and use cases, where data integrity and trust are paramount.

Creating a ledger in Azure SQL database

To create a ledger database in Azure SQL Database with ledger capabilities enabled at the database level, we need to use the CREATE DATABASE statement with the LEDGER = ON option. This feature enhances the database with cryptographic verifiability, allowing us to maintain an immutable record of data modifications.

First, let's connect to an instance of Azure SQL Server, either using **SQL Server Management Studio** (**SSMS**), Azure Data Studio, or any SQL client where it's possible to execute T-SQL commands. Then, we must create a new database by executing the following command:

```
CREATE DATABASE <database_name>
WITH LEDGER = ON;
```

Replace <database_name> with the name you wish to give your ledger-enabled database.

Once the database has been created, we can start using it immediately. We can create tables and perform CRUD operations as usual. The ledger functionality will be automatically integrated into the database, ensuring that all data modifications are recorded immutably.

To take advantage of ledger capabilities, such as verifying the integrity of the data or auditing changes, we'll need to familiarize ourselves with specific T-SQL extensions provided by Azure for ledger operations. These might include functions and procedures for cryptographic verification of the data stored in the ledger.

Database ledger versus table ledger

Creating a ledger-enabled database in Azure SQL and creating ledger-enabled tables within an Azure SQL database are two distinct operations that cater to different levels of data integrity and auditing requirements.

As we've seen previously, the `CREATE DATABASE WITH LEDGER = ON` command creates a ledger-enabled database. This command is used at the database level, and it enables ledger functionalities for the entire database. This means that the ledger capabilities, which include cryptographic verification of data integrity and immutability, can be used by any table within the database, but it's up to individual tables to utilize these features.

The primary goal here is to establish a database environment that supports ledger functionalities across the board, making it easier to manage and apply these capabilities on a table-by-table basis as needed. This is ideal for new databases where we anticipate the need for ledger functionalities across multiple tables, providing a foundation for advanced data integrity and auditing capabilities from the start.

Azure SQL Database also supports updatable ledger tables on which users can perform update and delete operations while also providing tamper-evidence capabilities. This type of ledger can be enabled with the `CREATE TABLE WITH LEDGER = ON` command.

This command is specific to a single table within a database. We can use it when creating a new table or altering an existing one to enable ledger functionalities for that specific table. It allows for the cryptographic verification of the data within the table, ensuring that all data modifications are recorded immutably.

The focus here is on applying ledger capabilities to individual tables. This is useful for databases that only require ledger functionalities for specific tables rather than at the entire database level. This is ideal for scenarios where only certain critical or sensitive data tables require the enhanced integrity and audit capabilities provided by ledger functionalities. It allows for a more granular application of these features, enabling them only where necessary.

Benefits of the ledger features of Azure SQL

Before we progress any further with technical details, let's dig into the specific benefits of the ledger features of Azure SQL, as identified in the previous section. Each benefit will be described in terms of impact and configuration or coding details for implementation.

Security and integrity

The ledger features of Azure SQL Database offer significant benefits concerning data security and integrity. By leveraging these features, organizations can ensure that their data remains tamper-proof, transparent, and resistant to unauthorized modifications. First, we'll explore how the ledger features enhance security and integrity, after which we'll dive into some code examples demonstrating their usage:

- **Data immutability**: The ledger features ensure that once a transaction is committed to the ledger, it becomes immutable. This means that historical data remains intact and cannot be altered or deleted, providing a reliable audit trail for all database changes. This immutability enhances data integrity and eliminates the risk of accidental or malicious modifications.

- **Cryptographic hashing**: Ledger features utilize cryptographic hashing algorithms to generate a unique hash for each transaction. These hashes create a chain of blocks, and any change to the data results in a new block with a new hash. This mechanism ensures that any attempt to tamper with the data will be evident in the form of a mismatched hash, signaling potential data manipulation.

- **Transparent transaction history**: Azure SQL Database's ledger features maintain a transparent and tamper-evident transaction history. Every change made to the database is recorded in the ledger, including the date, time, and details of the transaction. This transparency enables auditors and administrators to easily track the sequence of events leading to any specific data state, fostering trust and accountability.

- **Auditing made simple**: The ledger features simplify the auditing process. Organizations can efficiently query the transaction history to retrieve a full record of all data changes, making it easier to comply with regulatory requirements and internal audit needs.

Next, we'll provide some code examples to illustrate how to use the ledger features in Azure SQL Database to ensure improved data security and integrity.

Enabling a ledger on a table

In this example, we'll create a new table and enable the ledger on it. Once the ledger has been enabled, any data changes that are made to the table will be recorded in the transaction history.

The CREATE TABLE command defines the table structure, and the additional WITH LEDGER option will add an append-only ledger table. As the name implies, append-only ledgers only allow data to be inserted; it can never be updated or deleted. This is indicated by the APPEND_ONLY parameter. For scenarios where data updates are necessary, the APPEND_ONLY parameter can be omitted, and the table will be created with an updatable ledger. In this case, data is still versioned in the ledger, and updates and deletes can still occur. When an update or delete occurs, all rows preceding the current one being updated/deleted are preserved in a history table, and the latest version remains in the ledger:

```
-- Create a new table
CREATE TABLE Sales (
    ID INT PRIMARY KEY,
    ProductName NVARCHAR(50),
    Quantity INT,
    Price DECIMAL(10, 2)
) WITH (LEDGER = ON (APPEND_ONLY = ON));
```

Whenever a data operation occurs, the ledger table will undertake extra steps to keep the historical data up to date and calculate the digests recorded in the database ledger. In detail, for each row that is updated, the process involves the following aspects:

1. Saving the previous **version** of the row in a history table.

2. Assigning a **transaction ID** and creating a new sequence number, then recording these in designated system columns.

3. Serializing the content of the row and incorporating it into the **hash** calculation for all rows affected by this transaction.

Ledger tables accommodate these requirements by modifying the **Data Manipulation Language (DML)** execution plans for insert, update, and delete operations. Specifically, for the new version of each row, the transaction ID and a new sequence number are applied. Following this, an operation within the query plan serializes the row's content and calculates its hash. This hash is then added to a Merkle tree, which is kept at the transaction level and includes the hashes of all the row versions that have been modified in this transaction for the ledger table. The Merkle tree's root hash represents all the modifications and deletions carried out by the transaction on this ledger table. When a transaction involves multiple tables, a distinct Merkle tree is created for each table.

Reading data from the ledger

Selecting data from a database ledger or a table ledger in Azure SQL involves using standard SQL queries, with some nuances depending on whether we're querying the ledger for current data or historical data.

When we have ledger-enabled tables, we interact with them just like any other table for most read operations. To select the current data from a ledger-enabled table, we can use the standard SELECT statement:

```
SELECT * FROM Sales;
```

This query returns the current state of the data in the table.

As we're interested in the historical data for audit or verification purposes, we need to query the ledger table as a system-versioned temporal table with history tracking enabled. This can easily be done by using the FOR SYSTEM_TIME clause, which gets historical data:

```
SELECT * FROM Sales
FOR SYSTEM_TIME AS OF '2024-01-01T00:00:00.0000000Z';
```

This query allows us to see the state of the data at a specific point in time or the changes over time.

If we're working with a ledger-enabled database, the process of selecting current data doesn't change: we use standard `SELECT` statements. However, querying cryptographic digests or proofs to verify data integrity involves more specialized operations, which might require accessing specific system tables or using built-in functions provided by Azure SQL for ledger databases.

To view all the transactions in a database ledger, we can use the following query. The `database_ledger_transactions` object contains rows with the information of each transaction in the ledger. The data includes the block's ID, where this transaction occurred, and the sequence of the transaction within the block:

```
SELECT * FROM sys.database_ledger_transactions;
```

To get a list of all the blocks in a ledger, we can query the `database_ledger_blocks` object. Here, we can find all the blocks in the ledger, including the root of the Merkle tree, and the hash of the previous block, similar to what happens in a blockchain:

```
SELECT * FROM sys.database_ledger_blocks;
```

With these queries in mind, let's discuss the performance and scalability of data and transactions in the ledger.

Performance and scalability

Performance and scalability are essential aspects of any database system, and the ledger features in Azure SQL Database are designed to ensure that data integrity and transparency do not come at the cost of reduced performance or limited scalability. Leveraging ledger technology, organizations can benefit from the following aspects related to performance and scalability:

- **Efficient storage**: The ledger features use a technique called temporal data storage to efficiently manage historical records. Rather than creating full copies of each record for every transaction, temporal storage keeps track of only the changes made to the data. This approach minimizes storage requirements and allows for the retention of historical data without the need to consume excessive space.

- **Optimized transaction processing**: Azure SQL Database's ledger features are designed to handle high volumes of transactions efficiently. The underlying mechanisms for maintaining the ledger ensure that transaction processing remains performant, even as the volume of data and transactions grows.

- **Scalability with business growth**: As the volume of data and the number of transactions increase with business growth, the ledger features can scale accordingly. Azure SQL Database's scalability capabilities enable organizations to handle growing workloads without sacrificing performance.

- **Minimal impact on application performance**: The ledger features operate transparently in the background, which means that application performance is not significantly affected. The ledger is updated automatically as part of the underlying transaction processing, ensuring a seamless experience for application users.

Optimizing performance in a database ledger, especially within Azure SQL Database with ledger features enabled, involves balancing the integrity and immutability requirements of the ledger with the performance characteristics of the database. Let's dive into several strategies and examples to help optimize performance.

Indexing

Proper indexing is crucial for improving query performance. For ledger-enabled tables, consider indexes on frequently queried columns, including system-versioned columns if applicable. For example, we could add an index to the `ProductName` column in the `Sales` table:

```
CREATE NONCLUSTERED INDEX IX_Sales_ProductName
ON Sales (ProductName);
```

Modern databases can optimize the maintenance of indices on tables, although a good database administrator should always know these basic procedures for performance optimization.

Partitioning

Partitioning can help with managing large ledger tables by dividing them into smaller, more manageable pieces, based on a key. First, we need to define a partition function and scheme, as shown in the following example:

```
-- Create a partition function
CREATE PARTITION FUNCTION SalesPartitionFunctionName (int)
AS RANGE LEFT FOR VALUES (1, 2, 3);

-- Create a partition scheme
CREATE PARTITION SCHEME SalesPartitionScheme
AS PARTITION SalesPartitionFunctionName
ALL TO ([PRIMARY]);
```

Then, we must create or alter a ledger-enabled table to use the partition scheme. Now, this can be very simple if we're creating the table for the first time. In that case, we can just specify the partition in which the table will reside with the `CREATE TABLE ON <PartitionName>` command.

However, when a table already exists, the process is slightly longer. The process generally requires creating a new partitioned table, transferring data from the existing table to the new one, and then renaming the tables if necessary. Direct alteration of a table to add partitioning is not supported; you must recreate the table with the desired partition scheme. With the partition we created previously, let's create a second `Sales2` table. This new table will match the schema of the existing `Sales` table, including ledger and system-versioning columns if applicable, but it will specify the partition scheme for the primary key or clustered index:

```
CREATE TABLE Sales2 (
    -- Recreate all columns from the original Sales table
```

```
    PRIMARY KEY CLUSTERED (Id) ON SalesPartitionScheme (ProductName)
);
```

Ensure the partition key column (in this case, `ProductName`) is part of the primary key or clustered index, and use the partition scheme in the `ON` clause.

Now that we have the second table, we can migrate data from the original `Sales` table into the new one:

```
INSERT INTO Sales2 (Id, ProductName, ...)
SELECT Id, ProductName, ...
FROM Sales;
```

Once the data transfer is complete, we can rename the original table and then rename the new table so that it uses the original table's name. Please note that to be able to rename a ledger table, the `ALTER LEDGER` permission is required:

```
-- Rename original table (backup or remove later as needed)
EXEC sp_rename 'Sales', 'Sales_Old';

-- Rename new table to original name
EXEC sp_rename 'Sales2', 'Sales';
```

Additional good practices for performance optimization include writing queries that select only the required data or the use of query hints. Regularly maintaining the database is also another critical factor for the overall well-being of the database server. These tasks include updating statistics, checking index fragmentation, and performing consistency checks.

Lastly, for high-transaction scenarios, consider using memory-optimized tables for portions of the ledger data that are accessed or modified frequently.

Using the ledger features of Azure SQL for blockchain solutions

While the ledger features of Azure SQL Database are not full-fledged blockchain solutions like public blockchain platforms are, they can be utilized to implement certain aspects of blockchain-like functionality for specific use cases. Let's explore some practical examples of how the ledger features in Azure SQL Database can be used in blockchain solutions.

Supply chain traceability

In a supply chain solution, the ledger features can be employed to record and track the movement of goods and products from their source to the end consumer. Each transaction involving the product's movement, such as production, shipping, and delivery, can be recorded in the ledger. This ensures an immutable and transparent history of the product's journey through the supply chain, providing traceability and accountability.

The architecture diagram for this solution will consist of the following components:

- **Web application or user interface**: This component represents the user-facing interface where supply chain stakeholders interact with the system. It allows users to view product details, track shipments, and access historical data.

- **Azure SQL Database**: This component represents Azure SQL Database, where the supply chain transaction data is stored. The ledger features are enabled on this database to maintain a transparent and immutable transaction history.

- **Application backend/APIs**: The application backend or APIs facilitate communication between the web application and Azure SQL Database. It handles user requests, executes queries, and processes data.

- **Azure Active Directory (Azure AD – optional)**: Azure AD can be used for authentication and access control, ensuring secure access to the web application and database.

- **Azure App Service (optional)**: If the web application is hosted on Azure, you can use Azure App Service to deploy and manage the application.

The connections between these components in the following architecture diagram illustrate how data flows from IoT devices to the ledger in an Azure SQL Database, and how smart contracts on a blockchain verify the transaction data of the supply chain in real time:

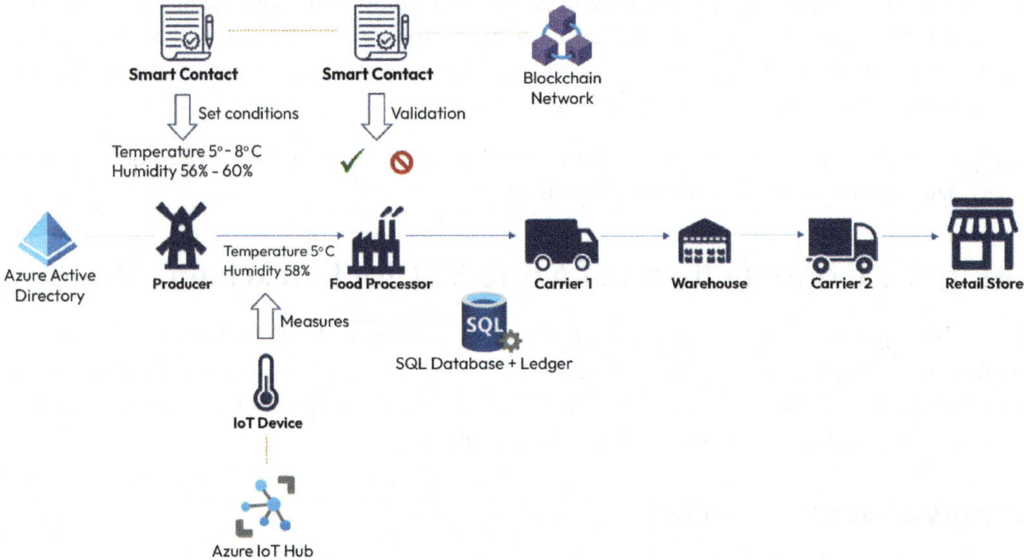

Figure 8.1 – Architecture diagram for a supply chain traceability solution

A simplified implementation for the supply chain solution when using the ledger features in Azure SQL Database may look like this:

```
-- Create a table for supply chain transactions
CREATE TABLE SupplyChainTransactions (
    TransactionID INT PRIMARY KEY,
    ProductID INT,
    Action NVARCHAR(100) NOT NULL,
    Location NVARCHAR(100),
    Timestamp DATETIME
) WITH LEDGER = ON APPEND_ONLY = ON;
```

In this example, the `SupplyChainTransactions` table stores the real-time supply chain transaction data, including details such as product ID, action (for example, production, shipping, or delivery), location, and timestamp.

The ledger table associated with it acts as the history table for the ledger features. It stores the historical records of supply chain transactions with their start and end timestamps, ensuring data immutability and transparency.

Certificate issuance and verification

For a certificate issuance and verification system, the ledger features can be utilized to maintain a transparent and tamper-proof record of issued certificates and their validity status. Each certificate issuance and revocation can be recorded in the ledger, ensuring the integrity of the certificate data.

A potential data model structure includes the `Certificates` table for storing information about issued certificates, and the associated ledger table, which has the ledger features enabled for traceability:

```
-- Create a table for certificates
CREATE TABLE Certificates (
    CertificateID INT PRIMARY KEY,
    HolderName NVARCHAR(100) NOT NULL,
    IssueDate DATE NOT NULL,
    ExpiryDate DATE,
    Status NVARCHAR(50)
) WITH LEDGER = ON APPEND_ONLY = ON;
```

The `Certificates` table holds a record of unique certificates that have been issued to a named holder, on a specific date, and with an optional (nullable) expiry date. Also, in this case, the ledger associated with this table keeps a record of all certificates being issued, to whom and when, in an immutable record.

Asset ownership transfer

In a system that deals with asset ownership transfer, such as real estate or intellectual property, the ledger features can be used to record and track ownership changes. Each transfer of ownership can be recorded in the ledger, providing a transparent and auditable trail of asset ownership history.

Also, in this case, our example of a data model includes two tables—`AssetOwnershipTransfers`, for storing asset ownership transfers, and the ledger-enabled history table:

```sql
-- Create a table for asset ownership transfers
CREATE TABLE AssetOwnershipTransfers (
    TransferID INT PRIMARY KEY,
    AssetID INT,
    PreviousOwner NVARCHAR(100),
    NewOwner NVARCHAR(100),
    TransferDate DATE
) WITH LEDGER = ON APPEND_ONLY = ON;
```

In these practical examples, we use the ledger features of Azure SQL Database to create transaction logs that serve as immutable records for specific blockchain-like use cases. While not a fully decentralized blockchain, this approach leverages the benefits of the ledger features, such as data immutability and transparent auditing, to enhance the security and trustworthiness of the recorded data. It is important to note that for more complex blockchain solutions that require decentralized consensus and public verification, a dedicated public blockchain platform would be more suitable.

Integrating the ledger features of Azure SQL with other Azure services

Complex scenarios will benefit from the tight integration of the ledger features of Azure SQL Database with other Azure services to build comprehensive and powerful solutions. Azure provides a wide range of services that can be combined with the ledger features to enhance data processing, analysis, and application capabilities. Let's explore some possible integration scenarios.

Azure Functions

We can integrate Azure Functions with the ledger features of Azure SQL Database to trigger serverless functions in response to data changes in the ledger. For example, when a specific event occurs in the ledger, such as a new product shipment being recorded in the supply chain, an Azure Function can be automatically triggered to notify relevant stakeholders via email, update a dashboard, or perform additional processing on the data.

Azure Logic Apps

Using Azure Logic Apps, we can create workflows that connect various services and applications. For instance, we can set up a Logic App to monitor the ledger transactions, and when specific conditions are met (for example, a high-value transaction), the Logic App can initiate a workflow to send notifications, generate reports, or execute other actions using different Azure services.

Azure Event Grid

Azure Event Grid enables event-driven architectures by providing a centralized event routing service. We can integrate the ledger features with Azure Event Grid to publish events whenever there are data changes in the Ledger. Subscribers to the events can then react accordingly, such as updating a cache, invoking a function, or sending notifications.

Azure Analysis Services

Azure Analysis Services allows us to build data models, perform data analytics, and create business intelligence dashboards. By integrating the ledger data with Azure Analysis Services, we can gain valuable insights and visualize trends and patterns in the transaction history.

Azure Machine Learning

Integrating the ledger data with Azure Machine Learning enables us to perform predictive analytics, anomaly detection, and data-driven decision-making. For example, we can use machine learning models to identify potential supply chain disruptions based on historical ledger data.

Power BI

Power BI is a powerful data visualization tool in the Microsoft ecosystem. We can connect Power BI to Azure SQL Database with ledger features to create interactive dashboards and reports that provide real-time insights into the supply chain or other ledger data.

Azure API Management

Azure API Management allows us to publish, manage, and secure APIs. By using API Management, we can expose certain functionalities of the ledger database as APIs, enabling external applications or partners to access specific ledger data securely.

Azure Data Factory

Azure Data Factory is a data integration service that can help orchestrate and automate data movement and transformation. We can use Data Factory to **Extract, Transform, and Load** (ETL) data from various sources into the ledger-enabled Azure SQL Database, creating a comprehensive data pipeline.

The combination of ledger technology and other Azure services enables us to build robust, scalable, and intelligent solutions tailored to our specific business needs and use cases. The possibilities for integration are extensive, and the choice of services depends on the requirements and objectives of a specific application or solution.

Best practices for implementing blockchain solutions with the ledger features of Azure SQL

Implementing blockchain solutions with the ledger features of Azure SQL Database requires careful planning and adherence to best practices to ensure the successful deployment and effective use of the technology. In this section, we'll cover some best practices to consider when implementing blockchain solutions with Azure SQL Ledger.

Initially, our solution should be configured to meet clear requirements for data consistency, security, and retention:

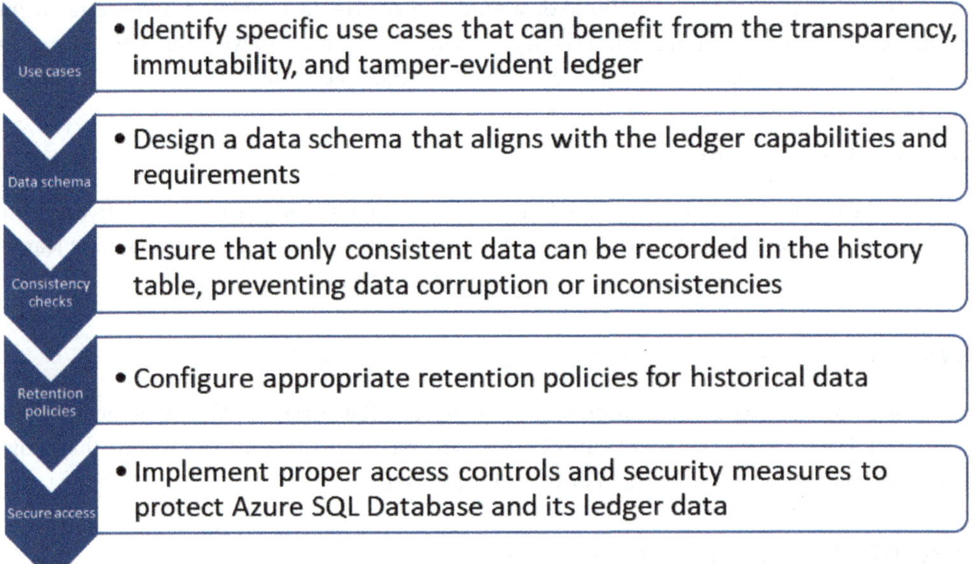

Figure 8.2 – Best practices for a secure and consistent implementation

When going through the stages of implementing a solution that leverages the ledger features, we may want to consider the following best practices:

- Identify specific **use cases** that can benefit from the transparency, immutability, and tamper-evident features of the ledger. Focus on scenarios where data integrity and auditability are critical, such as supply chain traceability, certificate issuance, or asset ownership tracking.

- Design a **data schema** that aligns with the ledger requirements. Ensure that the primary table contains all the necessary fields to record transactions, and create a history table for the ledger feature. Plan for appropriate fields to track transaction timestamps, previous and current values, and other relevant metadata.

- When enabling the ledger feature on a table, set the `DATA_CONSISTENCY_CHECK` option to `ON`. This ensures that only **consistent data** can be recorded in the history table, preventing data corruption or inconsistencies.

- Configure appropriate **retention policies** for historical data. Determine the retention period based on business needs and regulatory requirements. Setting a suitable retention period ensures that historical data remains accessible without burdening the database with unnecessary records.

- Implement proper **access controls** and security measures to protect Azure SQL Database and its ledger data. Use Azure Entra ID integration for authentication and **Role-Based Access Control (RBAC)** to restrict access to authorized personnel only.

Once the solution has been deployed, the following recommendations apply to runtime monitoring and optimization:

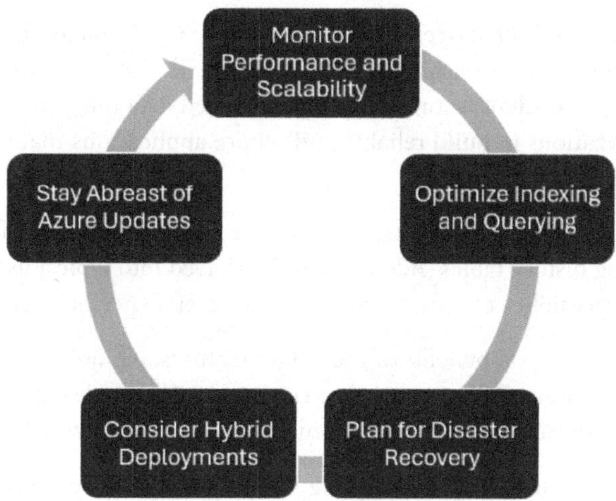

Figure 8.3 – Recommendations for monitoring and optimization

In detail, the recommendations are as follows:

- **Monitor** the performance of the database, especially as the number of transactions and historical records grows. Keep an eye on query performance, storage usage, and other resource metrics to ensure scalability as the blockchain solution expands.

- Choose appropriate **indexing** strategies to optimize query performance on the ledger-enabled tables. Use indexing on fields that are commonly used in filtering or sorting historical data to enhance query efficiency.

- Implement **disaster recovery** plans to ensure the safety of historical data in the ledger in case of any unforeseen incidents. Consider using Azure SQL Database's backup and restore features to maintain data resilience.

- Evaluate if a **hybrid deployment** of public blockchain technology and the ledger features of Azure SQL Database can provide a more comprehensive solution. Public blockchain platforms can offer decentralized consensus and public verification, complementing the data integrity provided by the ledger features.

- Stay informed about the **latest updates** and improvements to the ledger features in Azure SQL Database. Regularly review the Azure SQL Database documentation and announcements to be able to leverage new capabilities and enhancements.

By following these best practices, organizations can effectively leverage the ledger features of Azure SQL Database to implement blockchain-like functionality in their solutions while maintaining data integrity, transparency, and security.

Summary

In this chapter, we explored the ledger features of Azure SQL Database and their potential for implementing blockchain-like functionality in various solutions. The ledger features provide transparency, immutability, and tamper-evident records, offering enhanced data integrity and auditability. These features enable organizations to build reliable and secure applications that require a trustworthy historical record of data changes.

We also discussed the technical prerequisites for using the ledger features, including enabling system versioning and creating history tables. Additionally, we delved into typical use cases for the ledger, such as supply chain traceability, certificate issuance, and asset ownership tracking.

This chapter highlighted several benefits of the ledger features, including improved data security, simplified auditing, data immutability, and performance scalability. Code examples and configuration settings demonstrated how to implement these benefits effectively in Azure SQL Database.

In the next chapter, our journey of discovering the ledger-based storage capabilities in Azure will be enriched with Confidential Ledger, a standalone digital ledger service that's secured with hardware-protected enclaves.

Further reading

- Documentation about ledger for SQL Server 2022 and Azure SQL Database: `https://learn.microsoft.com/en-us/sql/relational-databases/security/ledger/ledger-landing-sql-server`

- Azure documentation on how to enable and use temporal tables (ledger) in Azure SQL Database: `https://learn.microsoft.com/en-us/azure/azure-sql/temporal-tables?view=azuresql`

9

Leveraging Azure Confidential Ledger

In an increasingly digital and interconnected world, data security has become a paramount concern for businesses and organizations of all sizes. The advent of cloud computing has revolutionized the way data is stored and processed, but it has also introduced new challenges in protecting sensitive information from unauthorized access and tampering. To address these concerns, Microsoft Azure, a leading cloud services platform, has developed **Azure Confidential Ledger** (**ACL**), a groundbreaking solution that sets new standards for data privacy, security, and integrity.

In this chapter, we'll dive into the following main topics:

- An introduction to ACL
- The features and benefits of ACL
- Using ACL for blockchain solutions
- Integrating ACL with other Azure services
- Best practices for implementing blockchain solutions with ACL

Technical requirements

To run ACL, we need to meet certain technical requirements to ensure a secure and optimal environment for data protection.

The following are the key technical requirements:

- **Azure subscription**: First of all, we certainly need an active Azure subscription to access and utilize ACL services. From the Azure portal, we will then be able to create a new instance of ACL by browsing Azure Marketplace. The following screenshot shows the **Confidential Ledger** component in the Azure portal:

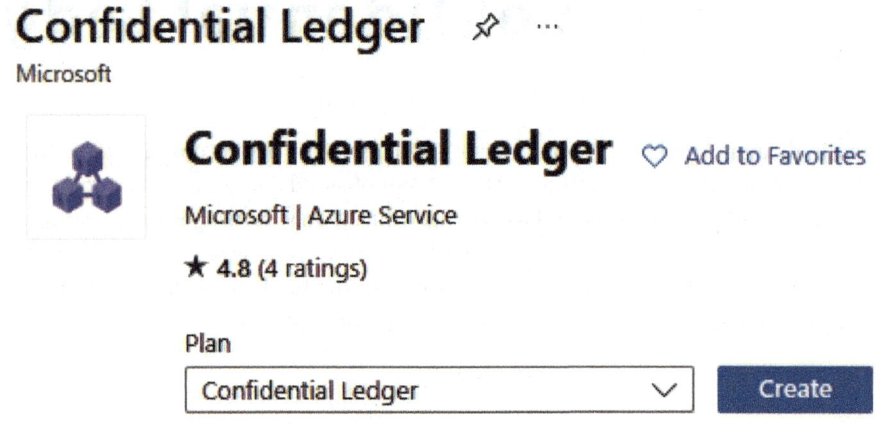

Figure 9.1 – The Confidential Ledger component in the Azure portal

- **Azure Entra ID**: Azure Entra ID is required for managing access and permissions to Confidential Ledger resources. It provides identity and access management capabilities, ensuring that only authorized users and applications can interact with the ledger. The following screenshot shows the **Security** settings of ACL in the Azure portal, where you can select whether you prefer an Azure Entra ID-based authentication or a certificate-based authentication. An alternative approach to using Azure Entra ID is to opt for certificate-based credentials:

Create confidential ledger or managed CCF   ···

Basics   **Security**   Tags   Review + create

Confidential Ledger: Users receive permissions to the ledger you create. You can add AAD or certificate users through this page. Assign Reader role for read-only permissions to the ledger, Contributor role for read/write permissions, and Administrator role for read/write and user management permissions.

Managed CCF: The member will receive permissions to perform governance operations like updating the constitution and managing members besides others. Learn more

+ Add AAD-Based User   🗑 Delete

	Name	Email	Ledger Role
☐	Stefano Tempesta		Administrator

+ Add Certificate-Based User   🗑 Delete

	Certificate	Ledger Role

Figure 9.2 – Security settings for ACL in Azure

- **Azure Key Vault**: To enhance security and manage cryptographic keys used for data encryption, we should utilize Azure Key Vault. This allows us to control access to keys and secrets, safeguarding them from unauthorized access.

- **Azure Private Link (optional)**: While not strictly required, Azure Private Link can be used to further secure network communications by granting access to Confidential Ledger over a private connection instead of the public internet.

Before deploying ACL, thoroughly review Microsoft's official documentation, best practices, and guidelines to ensure a secure and well-configured environment that meets specific application needs.

An introduction to ACL

ACL represents a significant step forward in the realm of confidential computing. Unlike traditional cloud data storage, where data may be accessible to cloud service providers and their infrastructure, Confidential Ledger takes data protection to a higher level. This innovative service allows organizations to secure their most critical and sensitive data while maintaining the benefits of cloud scalability, availability, and cost-effectiveness.

The core principle of ACL lies in the concept of **confidential computing**. It leverages **Trusted Execution Environments (TEEs)** to protect data and code from unauthorized access, even from the cloud provider itself. This ensures that data remains encrypted and is only processed in secure enclaves, thus shielding it from any potential breaches or attacks that may occur in the cloud environment.

One of the key features that sets ACL apart is its tamper-resistant nature. Data stored within the ledger is immutable, meaning once it is recorded, it cannot be altered or deleted. This capability not only provides data integrity but also establishes an auditable record of all transactions, enabling organizations to maintain a reliable and transparent data history. The following diagram shows the connections between client applications that send data to the ledger and the technical foundation of ACL, consisting of a distributed infrastructure based on hardware-secured enclaves:

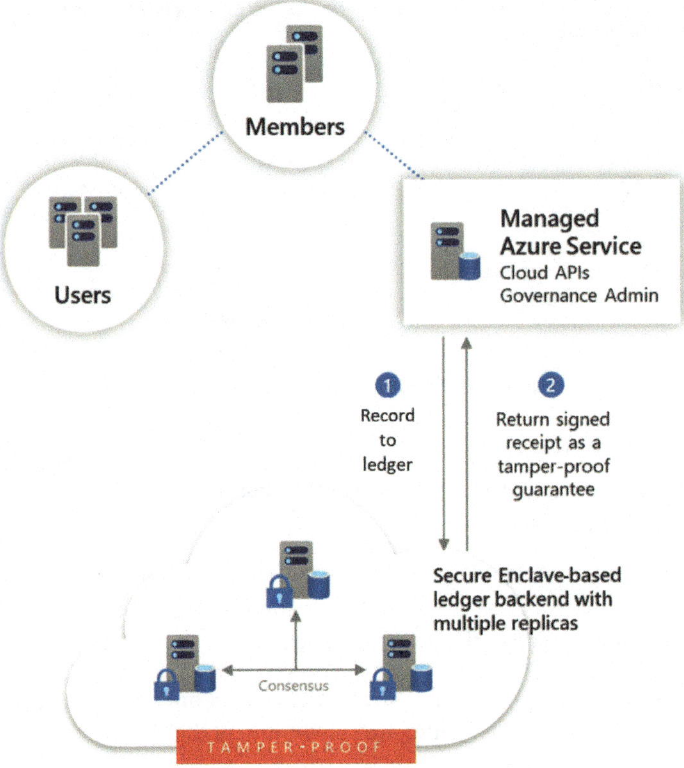

Figure 9.3 – A client-server representation of ACL

In this chapter, we will delve deeper into the architecture and working principles of ACL. We will explore how it empowers organizations to build applications with enhanced security and confidentiality, as well as the potential use cases across various industries. Additionally, we will examine the practical aspects of implementing Confidential Ledger, including integration with existing applications and development frameworks.

In conclusion, ACL represents a groundbreaking innovation that addresses the critical need for data privacy and security in the cloud. By leveraging confidential computing capabilities, organizations can now trust that their sensitive information remains safe and protected, even in a cloud environment. As we embark on this journey into the realm of confidential computing, let us uncover the vast possibilities and advantages that ACL brings to modern data management and security.

Use cases

ACL offers enhanced data security and confidentiality in the cloud by leveraging confidential computing capabilities. It is suitable for a wide range of use cases where data privacy, integrity, and immutability are critical.

Some common use cases of ACL include the following:

- **Supply chain management**: In supply chain management, ACL can be used to record and track the movement of goods, ensuring the authenticity and integrity of each transaction while protecting sensitive supply chain data.

- **Healthcare**: Confidential Ledger can securely store and manage patient health records, medical data, and clinical trial information, ensuring patient privacy and compliance with healthcare regulations.

- **Financial services**: In the financial sector, ACL can be used to securely manage and track financial transactions, such as trade settlements, fund transfers, and auditing, while keeping sensitive financial data confidential.

- **Intellectual property protection**: For industries dealing with intellectual property, such as software development and research, ACL provides a secure environment to store sensitive code, designs, and patents.

- **Blockchain and cryptocurrency**: ACL can complement existing blockchain solutions by providing a tamper-resistant and confidential storage layer for private data within a public or consortium blockchain network.

- **Legal and compliance records**: In legal and compliance use cases, Confidential Ledger can be used to maintain an immutable record of legal agreements, contracts, and compliance-related documents.

- **Sensitive data processing**: ACL can be used to securely process sensitive data, such as **Personal Identification Numbers** (**PINs**) or passwords, within confidential enclaves to protect it from unauthorized access.

- **IoT data security**: For IoT applications, Confidential Ledger can be used to store and manage data from connected devices securely, ensuring the privacy of user data and maintaining data integrity.

- **Research and development**: Confidential Ledger can safeguard sensitive research and development data, such as experimental results and proprietary information, while enabling collaboration among researchers.

- **Government and public services**: In government and public service applications, ACL can protect citizen data and enable secure data sharing between different government entities.

- **Secure data sharing**: Confidential Ledger facilitates secure data sharing and collaboration between multiple organizations without exposing the underlying sensitive information.

- **Digital identity and access management**: For identity and access management systems, ACL can store and manage digital identities and access credentials securely, reducing the risk of identity-related attacks.

These are just a few examples, and the potential use cases for ACL continue to expand as organizations seek advanced data protection solutions. The versatility of Confidential Ledger allows it to address a broad range of data security challenges across various industries and applications.

The features and benefits of ACL

ACL comes with a range of features and benefits that elevate data security and confidentiality in the cloud.

Figure 9.4 – The key features of ACL

Let's explore the key features summarized in the preceding diagram:

- **Secure enclaves**: ACL leverages Intel **Software Guard Extensions (SGX)** to create a TEE. This secure enclave ensures that sensitive data and code are protected from unauthorized access, even from cloud service providers and infrastructure.

- **Immutable and tamper-resistant**: Data stored in ACL is immutable, meaning that once it is recorded, it cannot be altered or deleted. This ensures data integrity and creates an auditable, tamper-resistant record of all transactions, providing a reliable and transparent data history.

- **Client-side encryption**: Confidential Ledger enables client-side encryption of data, providing an additional layer of protection before data is transmitted to the ledger. This ensures that data remains encrypted even during transit.

- **Integration with Azure services**: Confidential Ledger can seamlessly integrate with other Azure services, such as Azure Key Vault, Azure Active Directory, and Azure Virtual Networks, allowing users to leverage existing security mechanisms and access controls.

- **Identity and access management**: With integration into Azure Active Directory, Confidential Ledger allows for robust identity and access management. Organizations can define granular access policies, ensuring that only authorized users and applications can interact with the ledger.

- **Scale and performance**: ACL is designed to deliver high performance and scalability. It can handle large-scale data and transaction processing efficiently, making it suitable for enterprise-grade applications.

By utilizing a TEE, Confidential Ledger ensures that sensitive data remains protected and confidential, mitigating the risk of data breaches and unauthorized access. This makes ACL perfectly suitable to meet public and private regulatory compliance. Confidential Ledger's tamper-resistant nature and strong security features aid governments and organizations in meeting regulatory compliance requirements, especially in industries with stringent data protection standards.

On top of strong security, the immutability of data in Confidential Ledger allows for a transparent and auditable transaction history, making it ideal for applications where data integrity and transparency are critical.

However, unlike a blockchain network, ACL retains the benefits of cloud computing, such as scalability, availability, and cost-effectiveness. Its deployment is **Platform as a Service (PaaS)**, which means that Azure takes care of all the aspects of running the infrastructure, and you can focus on your business requirements. In addition, as an integral part of Azure's PaaS offerings, Confidential Ledger benefits from continuous updates and improvements, ensuring that it stays at the forefront of data security advancements.

ACL is developer-friendly too. It provides developers with a REST API and programming language-specific SDKs that make it easier to integrate ACL with existing applications, thereby encouraging the adoption of advanced security measures in your solutions.

In summary, ACL brings a wealth of features and benefits to safeguard sensitive data in the cloud. With its TEE, immutability, and seamless integration with Azure services, organizations can confidently deploy critical applications with enhanced data security and confidentiality.

Using ACL for blockchain solutions

Let's get started with ACL! Ensure you have an active Azure subscription, and then access the Azure portal at `https://portal.azure.com/`. Then, follow these steps to create a new instance of ACL:

1. In the Azure portal, navigate to **Create a resource** and search for `Confidential Ledger` in Azure Marketplace.

2. Select the **Confidential Ledger** service from the search results, as already shown in *Figure 9.1*, and click the **Create** button.

3. Configure the **Basics** settings:

    A. Select the subscription and resource group where you want to deploy the service.

    B. Provide a unique name for your Confidential Ledger instance.

    C. Choose the Azure region where you want to host your Confidential Ledger (consider your data residency requirements).

4. Configure the **Security** settings. Define access policies and roles for users and applications to interact with the Confidential Ledger instance. This may involve integrating with Azure Active Directory for identity management.

5. Review and create:

    A. Double-check your configuration settings to ensure everything is as desired.

    B. Click **Create** to initiate the provisioning of the ACL.

The provisioning process may take a few minutes to complete. During this time, Azure will create and configure the necessary resources for your Confidential Ledger instance.

Once the deployment is successful, you can access Confidential Ledger through the Azure portal or programmatically, using appropriate Azure SDKs and APIs.

Remember that ACL is a powerful security feature, but it's essential to design and implement your application securely to make the most of its capabilities. Additionally, always refer to the latest official documentation and guidelines from Microsoft for the most up-to-date information on provisioning ACL.

Connecting and sending data to ACL

Using ACL for blockchain solutions can significantly enhance the security and confidentiality of sensitive data within the blockchain network. ACL provides a secure and tamper-resistant storage environment for blockchain data and ensures that data remains encrypted and accessible only to authorized parties.

Before connecting and sending data to ACL, you may first want to identify any sensitive information that you want or don't want to store on a permanent and immutable ledger. This could include **Personally Identifiable Information** (**PII**), financial data, intellectual property, or any other confidential information.

By using ACL for blockchain solutions, you can effectively protect sensitive data from unauthorized access, reduce the risk of data breaches, and maintain data integrity and transparency. This combination of blockchain technology and confidential computing provides a robust foundation for building secure and privacy-preserving blockchain applications across various industries.

Once ready, you can incorporate ACL as the storage layer for sensitive data within your blockchain network. This is typically achieved through appropriate APIs and SDKs provided by Azure. The following is a conceptual Python code example that uses the Azure SDK for Python to interact with ACL:

```python
from azure.identity import DefaultAzureCredential
from azure.confidentialledger import ConfidentialLedgerClient,
LedgerEntry


credential = DefaultAzureCredential()
ledger_name = "your-confidential-ledger-name"
ledger_client = ConfidentialLedgerClient(ledger_name=ledger_name,
credential=credential)


def write_to_ledger(data_to_write):
    try:
        data_bytes = data_to_write.encode()
        entry = LedgerEntry(data=data_bytes)
        ledger_client.write(entry)
    except Exception as e:
        print("Error writing to the ledger:", str(e))


def read_from_ledger():
    try:
        latest_entry = ledger_client.get_last_entry()
        if latest_entry:
            data = latest_entry.data.decode()
            print("Data from the ledger:", data)
        else:
            print("No data found in the ledger.")
    except Exception as e:
        print("Error reading from the ledger:", str(e))
```

The code will do the following:

- Import the necessary Azure SDK modules.

- Set up the Azure credentials.

- Initialize `ConfidentialLedgerClient`.

- Define the `write_to_ledger` function, which creates a `LedgerEntry` object used to write data to the ledger.

- Define the `read_from_ledger` function to read data from the ledger.

The full source code for interacting with ACL using Python, including access control and error handling, can be found on the GitHub repository for this chapter at `https://github.com/PacktPublishing/Developing-Blockchain-Solutions-in-the-Cloud/tree/main/Chapter9`.

For the most current and comprehensive code examples, I recommend referring to the official Azure SDK documentation and samples for your preferred programming language, which can be found on the Azure SDK GitHub repository: `https://github.com/Azure/azure-sdk`.

Integrating ACL with other Azure services

Yes, we can integrate ACL with other Azure services to enhance the security and functionality of our applications. Azure provides various mechanisms and APIs to enable seamless integration between different services. Here are some examples of how we can integrate ACL with other Azure services:

- **Azure Key Vault** is a cloud service used to securely store and manage cryptographic keys, secrets, and certificates. You can use Azure Key Vault to manage encryption keys used by ACL to protect sensitive data. This ensures that encryption keys are securely stored and never exposed directly to an application.

- **Azure Entra ID** provides identity and access management services. By integrating ACL with Entra ID, we can control access to the ledger resources based on user identities, groups, or roles. This helps ensure that only authorized users and applications can interact with the confidential data stored in the ledger.

- **Azure Monitor** and **Log Analytics** provide monitoring, logging, and analytics capabilities for Azure resources. You can use these services to gain insights into the usage and performance of your ACL instance, track access patterns, and detect any suspicious activities. Be mindful, though, of not storing any sensitive data in Log Analytics. This would compromise the confidentiality that ACL offers, as Log Analytics is not a confidential service secured in a TEE.

- We can set up event-driven workflows by integrating ACL with **Azure Event Grid** and **Azure Logic Apps**. For example, we can trigger specific actions or notifications when new data is written to the ledger or when certain conditions are met.

- **Azure Functions** allows us to run serverless code in response to events. We can use Azure Functions to process data from ACL or to trigger actions based on changes in the ledger. Again, be mindful of not sending sensitive data from ACL to an Azure function, as functions are not secured by confidential computing technology.

- For certain use cases, we might want to store associated metadata or non-sensitive data related to the confidential data in ACL. We can use **Azure Storage** services (such as Azure Blob Storage or Azure Table Storage) to store such data.

The integration possibilities are vast, and they depend on the specific requirements of the application we're creating. ACL is designed to be compatible with various Azure services, enabling developers to build secure and robust solutions by leveraging the features and capabilities of the Azure cloud ecosystem.

Integration with Azure Key Vault

Let's now examine a potential scenario where we integrate ACL with **Azure Key Vault** (**AKV**) to securely manage encryption keys and secrets used by the ledger. The following is a Python code example that demonstrates how to use AKV to store and retrieve an encryption key, which can be used to encrypt data before writing it to ACL. It is assumed that you have already set up AKV and have appropriate permissions to create and retrieve secrets:

```python
from azure.identity import DefaultAzureCredential
from azure.keyvault.secrets import SecretClient
from azure.confidentialledger import ConfidentialLedgerClient,
LedgerEntry


key_vault_url = "https://your-key-vault-url.vault.azure.net/"
secret_name = "your-secret-name"
credential = DefaultAzureCredential()
secret_client = SecretClient(vault_url=key_vault_url,
credential=credential)


def get_encryption_key():
    try:
        secret = secret_client.get_secret(secret_name)
        encryption_key = secret.value.decode()
        return encryption_key
    except Exception as e:
        print("Error retrieving the encryption key:", str(e))
        return None


def encrypt_data(data, encryption_key):
    encrypted_data = bytes([b1 ^ b2 for b1, b2 in zip(data,
encryption_key)])
    return encrypted_data
```

This code will do the following:

- Set up the Azure credentials to instantiate a new AKV client.

- Define the `get_encryption_key` function to get the encryption key from Key Vault.

- Define the `encrypt_data` function to encrypt data to send to the ledger.

Note that this code example uses a simple XOR encryption for demonstration purposes only. A XOR encryption algorithm is an example of additive symmetric encryption where the same key is used to both encrypt and decrypt a message. In a real-world scenario, you should use a strong encryption algorithm, such as AES, along with proper key management practices.

Additionally, ensure that you have the appropriate access policies and permissions set up in AKV to allow your application to retrieve the encryption key securely. Also, make sure that you handle the encryption and decryption of data securely in your application to maintain the confidentiality of your sensitive information.

Best practices for implementing blockchain solutions with ACL

Implementing blockchain solutions with ACL requires careful consideration of security, privacy, and performance aspects. The following diagram shows some of the best practices to keep in mind:

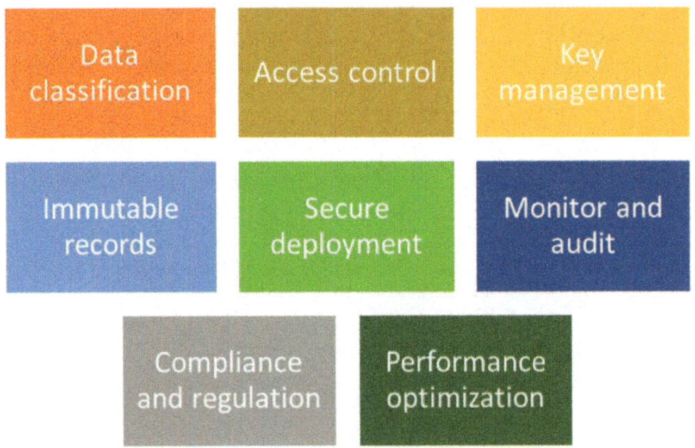

Figure 9.5 – The best practices for implementing blockchain solutions with ACL

Let's expand on each best practice:

- **Data classification**: Clearly identify and classify the data that requires confidentiality and protection. Use ACL only for storing sensitive and confidential data, while non-sensitive data can be stored in other components of the blockchain network. For example, when developing a blockchain solution, data may be written on chain – that is, on the blockchain itself – or off chain. In this case, ACL is a good storage option.

- **Access control**: Implement strict access control policies to regulate who can read, write, and modify data in ACL. Leverage **Azure Active Directory** (**Azure AD**) for identity and access management to ensure that only authorized users and applications can interact with the ledger.

- **Key management**: Use AKV to securely manage encryption keys used by ACL. Avoid hardcoding or exposing encryption keys directly in your code or configuration files.

- **Immutable records**: Take advantage of the immutability feature of ACL to maintain a tamper-resistant and auditable record of all transactions. This creates an accurate and transparent history of data changes.

- **Secure deployment**: Ensure that the blockchain network and ACL are deployed securely. Implement security best practices for both the blockchain platform and Azure services to reduce potential attack surfaces.

- **Monitor and audit**: Implement monitoring and auditing mechanisms to track access to ACL and detect any unusual activity or security breaches. Azure Monitor and Log Analytics can help with monitoring and logging.

- **Compliance and regulation**: Address compliance requirements and regulations related to data privacy and security. Confidential Ledger's capabilities can assist with meeting specific industry standards.

- **Performance optimization**: Optimize the performance of your blockchain solution with ACL. Consider factors such as ledger read/write throughput, network latency, and scalability to ensure efficient data processing.

By following these best practices, you can build a secure and privacy-preserving blockchain solution using ACL, protecting sensitive data while leveraging the benefits of blockchain technology and cloud services.

Summary

ACL is a groundbreaking solution by Microsoft Azure that sets new standards for data privacy, security, and integrity in the cloud. It falls under the umbrella of confidential computing and leverages TEEs to protect data and code from unauthorized access, even from cloud service providers themselves.

Overall, ACL provides a powerful solution for organizations seeking to secure sensitive data, maintain data integrity, and comply with strict data protection regulations in various industries. Its confidentiality capabilities complement existing blockchain solutions and enable developers to build secure and privacy-preserving applications in the cloud.

This chapter completes *Part 3* of the book, in which we have looked at three key blockchain-related services in the Azure cloud – Corda running on AKS, the ledger features of Azure SQL Database, and Confidential Ledger.

In the next chapter, we're starting *Part 4* of the book, which focuses on the blockchain services available in the **Google Cloud Platform (GCP)**.

Further reading

- For more information about ACL, the product page on Microsoft's website can be found at `https://azure.microsoft.com/products/azure-confidential-ledger`

- More specifically, ACL offers a fully REST API, documented at `https://learn.microsoft.com/rest/api/confidentialledger/`, as well as an SDK for Python, documented at `https://learn.microsoft.com/python/api/overview/azure/confidentialledger-readme?view=azure-python`

Part 4: Deploying and Implementing Blockchain Solutions on GCP

This part of the book will focus on deploying and implementing blockchain solutions on GCP. It will provide step-by-step instructions for building, deploying, and managing blockchain applications on Google Cloud Platform.

This part includes the following chapters:

- *Chapter 10, Hosting an Ethereum Blockchain Network on Google Cloud Platform*

- *Chapter 11, Getting Started with Blockchain Node Engine*

- *Chapter 12, Analyzing On-Chain Data with BigQuery*

10

Hosting an Ethereum Blockchain Network on Google Cloud Platform

In the rapidly evolving world of blockchain technology, Ethereum has emerged as one of the most prominent and widely adopted platforms for building decentralized applications and smart contracts. While Ethereum offers robust capabilities, deploying and managing a blockchain network can be a complex task. In its cloud platform, Google supports two deployment modes for Ethereum – one based on **Compute Engine** and another based on **Google Kubernetes Engine** (GKE), a fully managed Kubernetes service. Both offer robust solutions for deploying and managing an Ethereum network on containers, ensuring scalability, high availability, and fault tolerance.

In this chapter, we explore the seamless integration of the Ethereum blockchain with Compute Engine and GKE, delving into the steps, best practices, and benefits of hosting an Ethereum network on **Google Cloud Platform** (GCP). By harnessing the capabilities of both technologies, developers can unleash the full potential of their decentralized applications while enjoying the advantages of a cloud-native and scalable infrastructure.

We'll dive into the following main topics:

- Setting up an Ethereum blockchain network on Compute Engine
- Configuring nodes in the Ethereum network
- Managing the Ethereum network
- Troubleshooting and maintaining the Ethereum network

Technical requirements

To run the scripts presented in this chapter, you'll need a GCP account with active billing. If you don't have one already, you can create a new account in GCP by going to `https://cloud.google.com/`.

We will also need to enable the necessary APIs, including the Kubernetes Engine API, Compute Engine API, and Cloud Resource Manager API, as described later in this chapter, when setting up a GKE cluster.

Also, make sure you have the Google Cloud SDK installed and properly configured before running the scripts in this chapter. You can install the SDK from the official Google Cloud SDK documentation at `https://cloud.google.com/sdk/docs/install`

The complete source code for all the scripts presented in this chapter can be found in this book's GitHub repository: `https://github.com/PacktPublishing/Developing-Blockchain-Solutions-in-the-Cloud/tree/main/Chapter10`.

Setting up an Ethereum blockchain network on Compute Engine

In this section, we'll delve into the world of running an Ethereum network on Compute Engine. We'll explore the key components of an Ethereum network, the challenges involved in setting up a production-ready environment, and the benefits of leveraging Compute Engine's cloud-native features. By combining the power of Ethereum's decentralized architecture with the resilience of Compute Engine's infrastructure, developers can unlock the true potential of their blockchain applications while ensuring robustness, security, and seamless scalability.

In the next section, we'll look at how to deploy Ethereum on GKE, along with best practices for monitoring, managing, and securing your Ethereum network within GKE, enabling you to optimize performance, mitigate potential risks, and maintain the overall health of your blockchain infrastructure.

Whether you're an experienced blockchain developer or just starting your journey into decentralized applications, this chapter aims to equip you with the knowledge and tools needed to confidently run an Ethereum network on GCP.

Pointless to say, our journey starts by accessing the Google Cloud console at `https://console.cloud.google.com/`. Once you've signed into the console, search for `Ethereum` in the top search bar. The first entry in the list is the resource that we're going to use. It should look like this:

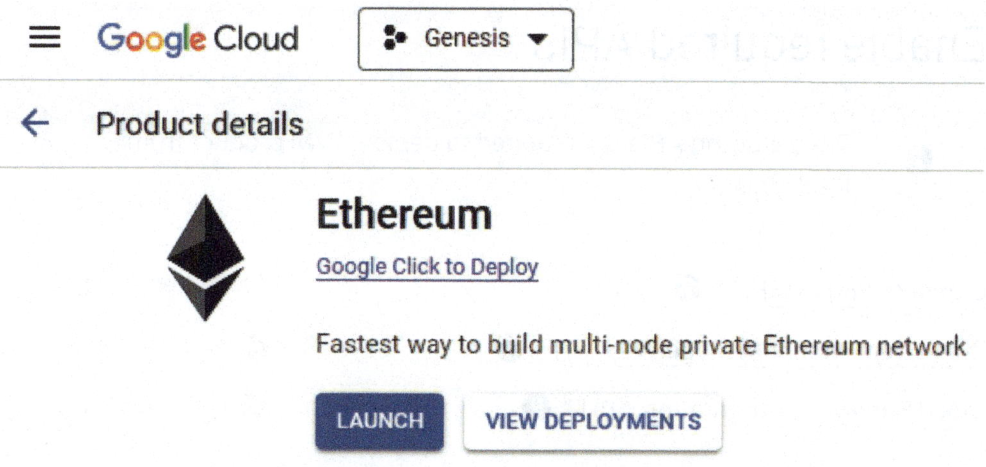

Figure 10.1 – Ethereum resource in GCP

This is the quickest way to deploy an Ethereum node on GCP, as we will see in the next section.

Let's continue with the Ethereum resource we've identified here. By using this templated approach, Ethereum will be deployed to a single Compute Engine instance with multiple Docker containers. We can then customize the configuration later when deploying the solution. Be mindful that Google does not offer support for this solution. However, community support is available on **Stack Overflow**. The software solution is open source, hence there's no license cost to pay. However, there is a cost associated with running the underlying VM instance, which can be estimated like so:

- **VM instance**: 1 vCPU + 3.75 GB of memory (n1-standard-1) about USD 35 per month

- **Standard persistent disk**: 10 GB of memory and about USD 0.50 per month

If we opt for a sustained-use discount – that is, VM instances are discounted up to 30% monthly based on their duration of use – we can save approximately USD 10 per month. Price estimates are based on 30-day, 24-hour per day usage of the listed resources in the Central US region.

Let's start the configuration process by selecting the **Launch** button. If this is the first time you're deploying a resource in your account, you'll be asked to enable the required APIs. Please do so before progressing to the next step:

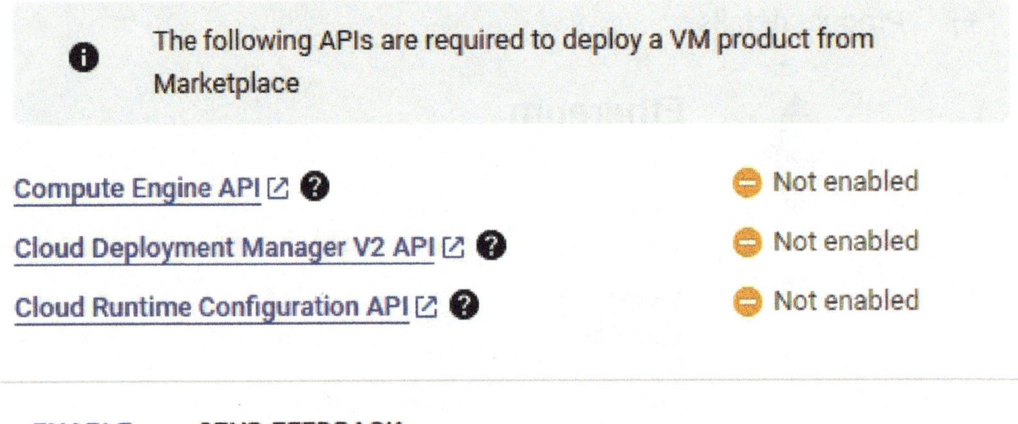

Figure 10.2 – Enabling the required APIs for deployment

Now, we must enter some information on the deployment screen:

1. First, you must enter a unique name that identifies your new Ethereum network and select the Google Cloud zone where you want to deploy it. This zone determines what computing resources are available and where your data is stored.

2. Then, select the machine type you wish to use, be it **General purpose** or **Compute-optimised**. **General purpose** specifies machine types for common workloads that have been optimized for cost and flexibility. **Compute-optimised** specifies machine types for performance-intensive workloads, with the highest performance per core. Since running a blockchain network is not a computing-intensive workload, I recommend choosing **General purpose**. Let's start with the **N1** series, which is powered by an Intel Skylake CPU, which has 1 vCPU and 3.75 GB of memory:

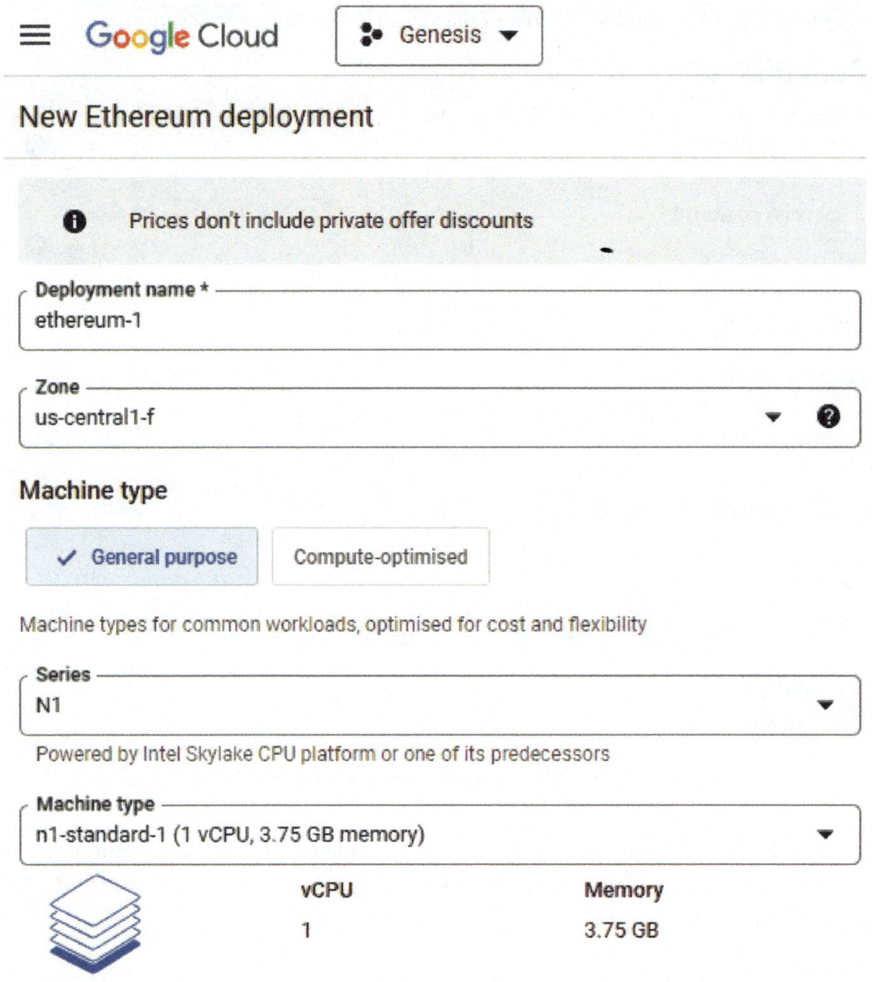

Figure 10.3 – New Ethereum deployment in Google Cloud

3. The third step is choosing the disk to attach to the VM. Although storage space is much less expensive for a standard persistent disk, I recommend selecting an SSD persistent disk, which offers better performance for IOPS throughput. A 10 GB size disk should suffice.

Accept the GCP Marketplace Terms of Service and then deploy this configuration:

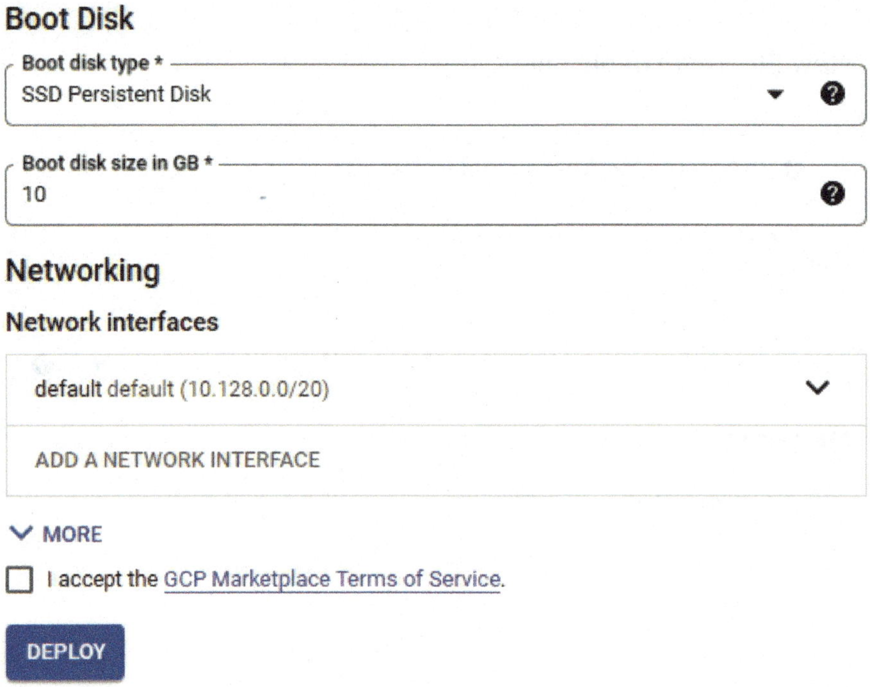

Figure 10.4 – Disk and networking settings

The deployment is going to take a few minutes. Once completed, you will have a full Ethereum node running on a Compute Engine VM in Google Cloud. All the infrastructure has been taken care of by Google as part of the deployment process.

Setting up an Ethereum blockchain network on GKE

The approach described previously is easy to execute and convenient to run, but it comes with a catch: our control over the deployed infrastructure is limited to the few options that we have seen before: zone, machine type, and disk size. How about all the aspects of container orchestration and scalability that Kubernetes offers? For this to happen, we need to deploy our Ethereum network in a different manner, which involves creating a Kubernetes cluster with GKE first and then deploying the relevant Ethereum nodes as pods.

Let's perform all these steps together:

1. **Create a Kubernetes cluster with GKE**

    Use the Google Cloud SDK to create a Kubernetes cluster with GKE. Choose the appropriate machine type, region, and node pool size based on your requirements:

    ```bash
    #!/bin/bash

    # Set your desired values
    PROJECT_ID="your-project-id"
    CLUSTER_NAME="your-cluster-name"
    ZONE="us-central1-a"
    NODE_COUNT=3
    MACHINE_TYPE="n1-standard-2"

    # Authenticate with your Google Cloud account
    gcloud auth login

    # Set the project
    gcloud config set project $PROJECT_ID

    # Create the GKE cluster
    gcloud container clusters create $CLUSTER_NAME \
        --zone $ZONE \
        --num-nodes $NODE_COUNT \
        --machine-type $MACHINE_TYPE

    # Get credentials for kubectl
    gcloud container clusters get-credentials $CLUSTER_NAME --zone $ZONE

    # Verify the cluster configuration
    kubectl config get-contexts
    ```

    This script does the following:

    - Authenticates you with your Google Cloud account

    - Sets the project to the specified project ID

    - Creates a GKE cluster with the given configuration (name, zone, node count, and machine type)

    - Retrieves cluster credentials and configures `kubectl` to use them

    - Verifies the cluster configuration

To execute it, please replace the placeholders (`your-project-id`, `your-cluster-name`, and so on) with your actual values. Save this script to a file (for example, `create_cluster.sh`), make it executable (`chmod +x create_cluster.sh`), and then run it (`./create_cluster.sh`) in your terminal.

2. **Install Ethereum nodes**

Once the GKE instance is ready, we can deploy the Ethereum nodes as Kubernetes pods in the cluster. We can use Ethereum client software such as Geth or Besu for that, and then configure the Ethereum nodes with the appropriate genesis file, network ID, and other necessary parameters.

We need to create a Kubernetes deployment and service for each node with the following YAML file to deploy a Geth Ethereum node. Save the following script in a `deployment.yaml` file:

```yaml
apiVersion: apps/v1
kind: Deployment
metadata:
  name: geth-node
spec:
  replicas: 1
  selector:
    matchLabels:
      app: geth-node
  template:
    metadata:
      labels:
        app: geth-node
    spec:
      containers:
      - name: geth-node
        image: ethereum/client-go:latest
        command: ["geth", "--datadir=/geth-data", "--rpc",
"--rpcapi=eth,net,web3,personal", "--rpccorsdomain=*"]
        ports:
        - containerPort: 8545
        volumeMounts:
        - name: geth-data
          mountPath: /geth-data
  volumeClaimTemplates:
  - metadata:
      name: geth-data
    spec:
      accessModes: [ "ReadWriteOnce" ]
      resources:
        requests:
          storage: 100Gi
```

```
---
apiVersion: v1
kind: Service
metadata:
  name: geth-node-service
spec:
  selector:
    app: geth-node
  ports:
  - port: 8545
    targetPort: 8545
```

3. **Configure a load balancer and ingress**

Now that we've covered the pods, we can set up a load balancer and ingress to manage external access to our Ethereum nodes. We will ensure that the load balancer is configured to route traffic to the appropriate Ethereum nodes. To do so, we will use the following YAML file:

```
apiVersion: v1
kind: Service
metadata:
  name: ethereum-load-balancer
spec:
  type: LoadBalancer
  ports:
  - port: 8545
    targetPort: 8545
  selector:
    app: geth-node
```

4. **Configure logging and monitoring**

At the end of the deployment, optionally but highly recommended, we can configure the logging and monitoring capabilities of the GKE cluster, auto-scalability, hardened security, and regular backups:

- To log and monitor an Ethereum network on GKE, I normally use tools such as **Prometheus** and **Grafana**

- The standard features available in Kubernetes and GKE for scaling the cluster size are normally sufficient

- To secure the Ethereum network, I would implement security measures such as node authentication, encryption, and firewalls to protect the Ethereum network from unauthorized access and attacks

- Finally, proper data backup and disaster recovery strategies will ensure data availability in case of unexpected failures

5. One final step that we may want to consider is introducing automation in our deployment pipeline by implementing **Continuous Integration and Continuous Deployment (CI/CD)**. For CI/CD in Google Cloud, I normally use **Google Cloud Build**, but you can use any other similar tool, such as **Jenkins** or **GitLab CI**. Here's an example of using Google Cloud Build:

```
# cloudbuild.yaml
steps:
- name: gcr.io/cloud-builders/docker
    args: ['build', '-t', 'gcr.io/$PROJECT_ID/my-ethereum-
app:$SHORT_SHA', '.']
- name: gcr.io/cloud-builders/kubectl
    args: ['apply', '-f', 'deployment.yaml']
```

This YAML file, which uses the project ID and cluster name defined in *Step 1* of the deployment process, simply automates the execution of the `deployment.yaml` file described before, which contains all the deployment instructions for a GKE cluster with Ethereum nodes.

Remember that deploying and managing an Ethereum network on GKE can be a complex process, especially if you are new to both technologies. It's essential to follow best practices, consult official documentation, and seek expert advice if needed. Additionally, ensure that you are aware of the costs associated with running your Ethereum network on GKE and monitor resource usage to optimize efficiency.

Configuring nodes in the Ethereum network

Configuring nodes in an Ethereum blockchain network involves setting up the Ethereum client software, connecting nodes to the network, and configuring their behavior. Nodes are essential participants in the blockchain network that are responsible for validating transactions, executing smart contracts, and propagating data across the network. Follow these steps to configure nodes in an Ethereum blockchain network:

1. **Install an Ethereum client software**

   Select an Ethereum client software that suits your requirements. Popular options include **Geth** (Go-Ethereum), **Besu** (Hyperledger Besu), **Nethermind**, and **EthereumJS**. Each client has its unique features and configurations.

   Install the chosen Ethereum client on each node. The installation process may vary, depending on the client and your operating system. The following instructions apply to Geth.

Save the YAML configuration in a file named `geth-deployment.yaml`:

```
apiVersion: apps/v1
kind: Deployment
metadata:
  name: geth-node
spec:
  replicas: 1
  selector:
    matchLabels:
      app: geth-node
  template:
    metadata:
      labels:
        app: geth-node
    spec:
      containers:
      - name: geth
        image: ethereum/client-go:v1.11.6   # Use the
appropriate Geth image version
        ports:
        - containerPort: 8545  # RPC port
        - containerPort: 30303 # P2P port
```

Apply the deployment to your GKE cluster using the `kubectl` command:

```
kubectl apply -f geth-deployment.yaml
```

This script creates a Kubernetes deployment named `geth-node` with a single replica. The deployment uses the Ethereum Geth Docker image and exposes two ports: `8545` for the RPC interface and `30303` for the P2P network interface. You can adjust the replica count and other configurations as needed.

After applying the deployment, Kubernetes will create a pod running the Geth container in your GKE cluster. You can then use Kubernetes services, load balancers, and network policies to manage and access the Geth node as necessary.

2. **Configure the genesis file (for new networks)**

As we're creating a new Ethereum network, we need to define the genesis block and network parameters in a JSON-based configuration file called the genesis file. The genesis block is the first block in the blockchain. The following example is a basic template for a Proof of Authority (Clique) private network. Save the content into a `genesis.json` file. We are going to use this file in the next step:

```
{
  "config": {
    "chainId": 1337,
```

```
    "homesteadBlock": 0,
    "eip150Block": 0,
    "eip155Block": 0,
    "eip158Block": 0,
    "byzantiumBlock": 0,
    "constantinopleBlock": 0,
    "petersburgBlock": 0,
    "istanbulBlock": 0,
    "clique": {
      "period": 15,
      "epoch": 30000
    }
  },
  "nonce": "0x0",
  "timestamp": "0x5E2E70C3",
  "extraData": "0x00000000000000000000000000000000000000000000
0000000000000000000",
  "gasLimit": "0x47B760",
  "difficulty": "0x1",
  "mixHash": "0x00000000000000000000000000000000000000000000000
0000000000000000",
  "coinbase": "0x0000000000000000000000000000000000000000",
  "alloc": {
    "address": {
      "balance": "1000000000000000000000000"
    }
  }
}
```

3. **Create an Ethereum data directory**

Before running the Ethereum client, we need to create a data directory for the blockchain data. This directory will store the blockchain database and other important files. This data directory is essential for running Ethereum clients such as Geth.

We also need to decide where we want to store our Ethereum data directory. This could be on our local machine, a server, or cloud storage. Let's see if we have enough storage space for the blockchain data, which can grow over time. Then, we can create a new directory to serve as our Ethereum data directory. I normally name it something like ethereum-data or geth-data.

As we're starting a new Ethereum node, we need to initialize it. This process involves creating the initial blockchain data. We will use the Ethereum client's commands to initialize the node. For example, with Geth, we can use the geth init command:

```
geth --datadir /path/to/ethereum-data init /path/to/genesis.json
```

Replace /path/to/ethereum-data with the actual path to your Ethereum data directory and /path/to/genesis.json with the path to your genesis configuration file.

Once the data directory is set up, we can start the Ethereum client using the --datadir flag to specify the location of the data directory. For example, with Geth, we can use the following command:

```
geth --datadir /path/to/ethereum-data
```

Again, replace /path/to/ethereum-data with the actual path to your Ethereum data directory.

When you start the Ethereum client for the first time, it will begin syncing with the Ethereum network to download the blockchain data. This process can take a significant amount of time, especially for the **mainnet**. Allow the client to complete the synchronization process.

4. **Connect nodes to the network**

To create a functioning Ethereum network, nodes need to be connected and communicate with each other. This can be achieved by specifying the **enode URLs** of other nodes in the network during node initialization. Assuming we're using the Geth client, we need to execute the following scripts to create a static nodes file, create a Kubernetes service for Geth nodes, configure Geth to connect to the static nodes, and then deploy the Kubernetes resources. Let's go through this in order.

In Ethereum, nodes can have a list of peers to connect to on startup. Create a JSON file named static-nodes.json with an array of Ethereum node URLs to connect to. Place this file in a location accessible to your GKE nodes:

```
[
    "enode://enode-url-1",
    "enode://enode-url-2",
    // ... add more enode URLs
]
```

We'll proceed by creating a Kubernetes service to expose the Geth nodes within the GKE cluster. This allows other nodes to discover and connect to our Geth nodes:

```
apiVersion: v1
kind: Service
metadata:
  name: geth-nodes
spec:
  selector:
    app: geth-node  # Match this to the label in your Geth
deployment
  ports:
```

```
      - protocol: TCP
        port: 30303  # P2P port
        targetPort: 30303
```

Then, we must modify our Geth deployment so that it includes the –bootnodes flag, pointing to our static nodes file. Also, we must use the Kubernetes service's DNS name to connect to peers:

```
apiVersion: apps/v1
kind: Deployment
metadata:
  name: geth-node
spec:
  # ... other settings
  template:
    spec:
      containers:
      - name: geth
        image: ethereum/client-go:v1.11.6    # Use your Geth
                                               image
        command: ["geth"]
        args: [
          "--datadir", "/ethereum-data",
          "--bootnodes", "enode://<enode-url>",
          "--networkid", "<network-id>",
          "--maxpeers", "25",
          "--port", "30303",
          "--rpc",
          "--rpcaddr", "0.0.0.0",
          "--rpcport", "8545",
          "--rpcvhosts", "*"
          # ... other flags
        ]
```

In this YAML file, we need to replace `<enode-url>` with an actual Ethereum node (enode) URL from our static nodes file and `<network-id>` with the Ethereum network ID.

Lastly, to deploy the Kubernetes resources to the GKE cluster, we must apply the Kubernetes service and deployment files with the `kubectl apply` command:

```
kubectl apply -f your-service-file.yaml
kubectl apply -f your-deployment-file.yaml
```

5. **RPC and Web3 API settings**

At some point, we will need to interact with our Ethereum nodes to deploy smart contracts. This requires us to configure the JSON-RPC and Web3 API endpoints, which enable external applications to communicate with the nodes and access blockchain data.

This step can be done at any time after the configuration of the Ethereum nodes is complete. Setting up RPC and the Web3 API in a Geth node running on GKE can be completed in a few steps:

I. **Modify the Geth deployment**: In our Geth deployment configuration YAML, we must add flags to enable RPC and the Web3 API. We also need to specify the allowed RPC origins to restrict access to our node:

```
apiVersion: apps/v1
kind: Deployment
metadata:
  name: geth-node
spec:
  # ... other settings
  template:
    spec:
      containers:
      - name: geth
        image: ethereum/client-go:v1.11.6    # Use your Geth
                                              image
        command: ["geth"]
        args: [
          "--datadir", "/ethereum-data",
          "--rpc",
          "--rpcaddr", "0.0.0.0",
          "--rpcport", "8545",
          "--rpcapi", "eth,web3",  # Enable specific APIs
          "--rpccorsdomain", "https://your-external-app.
          com"  # Allow CORS from your app
          # ... other flags
        ]
```

II. **Expose the RPC port using a service**: We need to create a Kubernetes service to expose the RPC port (8545) externally within our GKE cluster:

```
apiVersion: v1
kind: Service
metadata:
  name: geth-rpc
spec:
  selector:
    app: geth-node  # Match this to the label in your Geth
          deployment
  ports:
    - protocol: TCP
```

```
    port: 8545
    targetPort: 8545
```

III.   **Apply the service**: You can do this using the `kubectl apply` command:

```
kubectl apply -f your-service-file.yaml
```

> **Important – Secure the RPC interface**
>
> Exposing the RPC interface can be a security risk if not properly secured. Make sure you use strong authentication and restrict access to trusted IP addresses or domains. You can use additional tools such as reverse proxies or security groups to further enhance security.

6.   **Monitoring and metrics**

To monitor the health and performance of our Ethereum nodes, we may want to consider integrating monitoring and metrics tools such as **Prometheus** and **Grafana**. This is outside the scope of this chapter, but it's easy to configure. Please refer to the relevant documentation for details, such as *Google Cloud Managed Service for Prometheus*: `https://cloud.google.com/stackdriver/docs/managed-prometheus`.

7.   **Security considerations**

Last but not least, we will consider security measures to protect our nodes from unauthorized access and potential attacks. This includes firewall configurations, restricting RPC access, and keeping our Ethereum software up to date.

With that, you've configured the Ethereum nodes in GKE. In the next section, we'll look at best practices and techniques for managing the network when it's deployed at its best.

Managing the Ethereum network on GKE

Managing the Ethereum blockchain network on GKE involves various tasks to ensure the smooth operation, scalability, and reliability of the network. Here are the key aspects of managing an Ethereum network on GKE:

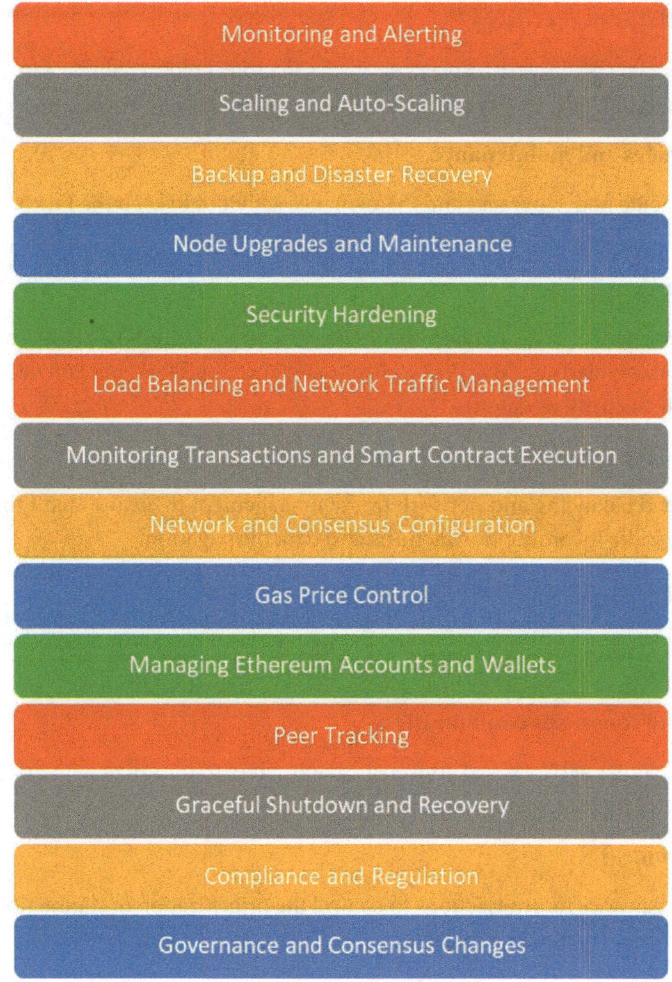

Figure 10.5 – Ethereum on GKE management stack

Let's elaborate some more on these key aspects:

- **Monitoring and alerting**

  Implement monitoring and alerting mechanisms to track the health and performance of the Ethereum nodes and the overall network. Use tools such as **Prometheus** and **Grafana** to monitor metrics and set up alerts for potential issues.

- **Scaling and auto-scaling**

  Set up auto-scaling for the Kubernetes cluster to automatically adjust the number of Ethereum nodes based on demand. This ensures the network can handle varying workloads efficiently.

- **Backup and disaster recovery**

  Establish backup and disaster recovery strategies to protect blockchain data in case of unexpected failures. Regularly back up the blockchain database and other critical data to prevent data loss.

- **Node upgrades and maintenance**

  Plan and execute node upgrades and maintenance activities to keep the Ethereum client software up to date and secure. Ensure seamless rolling upgrades to minimize disruptions.

- **Security hardening**

  Continuously monitor and manage the security of the Ethereum network. Implement security best practices, regularly audit access controls, and keep the Ethereum software and GKE cluster patched with security updates.

- **Load balancing and network traffic management**

  Optimize load balancing and network traffic management to ensure that Ethereum nodes can efficiently handle incoming transactions and data propagation.

- **Monitoring transactions and smart contract execution**

  Monitor the progress of transactions and smart contract execution on the network to identify potential bottlenecks or performance issues.

- **Network and consensus configuration**

  Review and adjust network and consensus parameters to optimize network performance and achieve the desired transaction throughput and block time.

- **Gas price control**

  Manage the gas price for transactions to control the prioritization of transactions and incentivize miners to include them in blocks.

- **Managing Ethereum accounts and wallets**

  Implement secure account management practices and use encrypted wallets to protect private keys.

- **Peer tracking**

  Keep track of connected peers and manage peer connections to ensure a healthy network topology.

- **Graceful shutdown and recovery**

  Plan and test graceful shutdown and recovery procedures to handle network disruptions or planned maintenance events.

- **Compliance and regulation**

  Ensure compliance with relevant regulations and standards, especially if the Ethereum network involves handling sensitive data or financial transactions.

- **Governance and consensus changes**

  If the Ethereum network is part of a consortium or private blockchain, plan and coordinate governance and consensus changes with network participants.

Remember that managing an Ethereum network is an ongoing process that requires continuous monitoring, maintenance, and optimization. Each aspect mentioned here may have several sub-tasks and configurations that need careful attention. In the next few sections, we are going to dive into more details regarding some of the most important tasks in the described management stack.

Auto-scaling

Scaling and auto-scaling in GKE can be achieved by configuring **Horizontal Pod Autoscaler (HPA)** for the deployed Ethereum nodes. HPA allows us to automatically scale the number of replicas (pods) based on CPU utilization or custom metrics. Here are the necessary scripts to enable scaling and auto-scaling for our Ethereum nodes on GKE:

1. **Enable the Kubernetes Metrics Server**

   First, we need to ensure that the Kubernetes Metrics Server is enabled in our GKE cluster. The Metrics Server provides resource utilization metrics for pods and nodes, which is essential for HPA to work. Run this bash script to enable the Metrics Server:

   ```
   # Enable the Metrics Server
   kubectl apply -f https://github.com/kubernetes-sigs/metrics-
   server/releases/latest/download/components.yaml
   ```

2. **Configure HPA**

   Next, we'll create an HPA for our Ethereum deployment to automatically adjust the number of replicas based on CPU utilization. This is the YAML script for configuring HPA:

   ```
   # ethereum-hpa.yaml
   apiVersion: autoscaling/v2beta2
   kind: HorizontalPodAutoscaler
   metadata:
     name: ethereum-hpa
   spec:
     scaleTargetRef:
       apiVersion: apps/v1
       kind: Deployment
       name: geth-node  # Replace with the name of your Ethereum
   deployment
     minReplicas: 1
     maxReplicas: 10  # Set the maximum number of replicas you want
   for auto-scaling
     metrics:
   ```

```
  - type: Resource
    resource:
      name: cpu
      target:
        type: Utilization
        averageUtilization: 70  # Set the average CPU
utilization threshold for scaling
```

3. **Apply the HPA configuration**

    In conclusion, we'll apply the HPA configuration to our Kubernetes cluster by running the following bash script:

    ```
    kubectl apply -f ethereum-hpa.yaml
    ```

With the HPA in place, Kubernetes will automatically adjust the number of replicas (Ethereum nodes) based on the CPU utilization of the pods. If the average CPU utilization exceeds the defined threshold (70% in this example), Kubernetes will scale up the number of replicas. If the utilization decreases below the threshold, Kubernetes will scale down the replicas accordingly.

Remember to adjust the HPA configuration based on your Ethereum deployment's resource requirements and your desired scaling behavior. It's also crucial to test the auto-scaling behavior in a controlled environment before deploying it in production.

> **A note on auto-scaling**
>
> The preceding example uses CPU utilization for auto-scaling, but we can also use custom metrics such as the Ethereum transaction count or block time for more accurate scaling based on specific Ethereum network requirements. To use custom metrics, we'll need to set up a custom metrics server and adapt the HPA configuration accordingly.

Load balancing

To set up load balancing for our Ethereum nodes, we can use Kubernetes services with a type of **load balancer**. This will create an external load balancer that can route traffic to the Ethereum nodes. Here are the necessary scripts to configure load balancing in GKE:

1. **Create a Kubernetes service**

    The following YAML script defines the parameters for creating a Kubernetes service that exposes the Ethereum nodes using a load balancer:

    ```
    # ethereum-service.yaml
    apiVersion: v1
    kind: Service
    metadata:
    ```

```
   name: ethereum-load-balancer
spec:
  selector:
    app: geth-node  # Replace with the labels of your Ethereum
deployment
  ports:
  - port: 8545  # The port on which Ethereum nodes listen for
incoming connections
    targetPort: 8545  # The port on which Ethereum nodes handle
the traffic
  type: LoadBalancer
```

2. **Apply the service configuration**

   With the YAML file we created previously, we'll apply the service configuration to the Kubernetes cluster by running the `kubectl apply` bash command:

   ```
   kubectl apply -f ethereum-service.yaml
   ```

   Kubernetes will automatically create a load balancer that will distribute incoming traffic to the Ethereum nodes labeled with `app: geth-node`, or any other label specified in the `selector` field.

   The external IP address of the load balancer will be allocated by GKE and will serve as the entry point to access the Ethereum nodes from outside the cluster.

   Keep in mind that it may take a few moments for GKE to allocate the external IP address for the load balancer.

3. **Access Ethereum nodes via the load balancer**

   Once the load balancer has been set up and the external IP address has been allocated, we can access the Ethereum nodes using the provided IP address and the specified port (in this case, port `8545`). The following bash script is an example of how to execute a `geth attach` command on the Ethereum nodes using the IP address of the load balancer:

   ```
   # Access Ethereum nodes using the external IP address of the
   Load Balancer
   # Replace EXTERNAL_IP with the actual external IP address
   geth attach http://EXTERNAL_IP:8545
   ```

Don't forget to secure access to the Ethereum nodes by configuring firewall rules and other access controls to restrict access to authorized entities only.

Gas price management

Gas price management in Ethereum involves configuring the gas price for transactions to control the prioritization of transactions and incentivize miners to include them in blocks. Here are the necessary scripts to manage the gas price in the Ethereum network deployed on GKE:

1. **Set a gas price in the Ethereum client**

    We can specify the default gas price for transactions in our Ethereum client's configuration with the **Geth** client using the `--gasprice` command-line option:

    ```
    geth --gasprice "20000000000"
    ```

    The indicated value is in Wei, the smallest denomination of Ether (1 Ether = 10^{18} Wei).

2. **Dynamic gas price**

    Instead of using a fixed gas price, we can implement a dynamic gas pricing mechanism. For example, we can use an external oracle or a custom algorithm to determine the gas price based on network congestion, transaction volume, or other factors.

    Here's an example of a simple bash script that uses a random gas price between a minimum and maximum range:

    ```bash
    #!/bin/bash
    MIN_GAS_PRICE=20000000000  # Minimum gas price in Wei
    MAX_GAS_PRICE=40000000000  # Maximum gas price in Wei


    # Generate a random gas price within the specified range
    GAS_PRICE=$(shuf -i $MIN_GAS_PRICE-$MAX_GAS_PRICE -n 1)


    # Start your Ethereum client with the dynamic gas price
    geth --gasprice "$GAS_PRICE" ...
    ```

    This script generates a random gas price between 20 GWei and 40 GWei (1 GWei = 10^{9} Wei) and passes it as the gas price when starting the Ethereum client.

3. **Transaction overrides**

    In some cases, we might want to set a custom gas price for specific transactions rather than using the default or dynamically calculated gas price. We can do this when sending transactions programmatically or interactively using tools such as **web3.js**.

    For example, using web3.js, we can specify the gas price in the transaction object, as shown in the following JavaScript, which sets a custom gas price of 30 GWei for the transaction:

    ```javascript
    const Web3 = require('web3');
    const web3 = new Web3('http://localhost:8545'); // Replace with
    your Ethereum client endpoint
    ```

```
const txObject = {
  from: '0xYourAddress',
  to: '0xRecipientAddress',
  value: web3.utils.toWei('1', 'ether'),
  gasPrice: web3.utils.toWei('30', 'gwei') // Custom gas price
in GWei
};

web3.eth.sendTransaction(txObject)
  .on('transactionHash', function(hash) {
    console.log('Transaction hash:', hash);
  })
  .on('confirmation', function(confirmationNumber, receipt) {
    console.log('Confirmation number:', confirmationNumber);
  })
  .on('receipt', function(receipt) {
    console.log('Receipt:', receipt);
  })
  .on('error', function(error) {
    console.error('Error:', error);
  });
```

Gas price management is an essential aspect of Ethereum network optimization. Depending on a case-by-case basis, we can choose a fixed, dynamic, or customized gas price strategy to balance transaction speed and cost in our Ethereum network. We should always consider factors such as network congestion and user preferences when determining the gas price policy for our application.

Managing Ethereum accounts and wallets

Managing Ethereum accounts and wallets involves creating, importing, and securely handling private keys to interact with the Ethereum network. In this section, we'll look at the necessary scripts to manage Ethereum accounts and wallets using the **web3.js** library, a popular JavaScript library for interacting with Ethereum:

1. **Install web3.js**

    First, we need to install the **web3.js** library using **npm**:

    ```
    npm install web3
    ```

2. **Create or import an Ethereum account**

    To create or import an Ethereum account, we can use the following JavaScript. Two methods are implemented in the script. The first method creates a new account using the `web3.eth.accounts.create()` function. The second method imports an existing account using its private key:

    ```
    const Web3 = require('web3');
    const web3 = new Web3('http://localhost:8545'); // Replace with
    your Ethereum client endpoint

    // Method 1: Create a new account
    const newAccount = web3.eth.accounts.create();
    console.log('New account address:', newAccount.address);
    console.log('New account private key:', newAccount.privateKey);

    // Method 2: Import an existing account using a private key
    const privateKey = '0x...'; // Replace with the private key of
    the account you want to import
    const importedAccount = web3.eth.accounts.
    privateKeyToAccount(privateKey);
    console.log('Imported account address:', importedAccount.
    address);
    console.log('Imported account private key:', importedAccount.
    privateKey);
    ```

3. **Securely store private keys**

    Handling private keys requires the utmost security since anyone with access to the private key can control the associated account. It's crucial to store private keys securely, ideally using a hardware wallet or a secure key management system.

4. **Sign transactions**

    To interact with the Ethereum network, we need to sign transactions using the account's private key. Here's an example of how to sign a transaction using the **web3.js** JavaScript library:

    ```
    const Web3 = require('web3');
    const web3 = new Web3('http://localhost:8545'); // Replace with
    your Ethereum client endpoint

    const accountAddress = '0x...'; // Replace with the address of
    the account you want to use
    const privateKey = '0x...'; // Replace with the private key of
    the account
    ```

```javascript
// Example transaction object
const txObject = {
  from: accountAddress,
  to: '0xRecipientAddress',
  value: web3.utils.toWei('1', 'ether'),
  gas: 21000, // Gas limit
  gasPrice: web3.utils.toWei('30', 'gwei') // Gas price in GWei
};

// Sign the transaction
web3.eth.accounts.signTransaction(txObject, privateKey)
  .then(signedTx => {
    // Send the signed transaction
    return web3.eth.sendSignedTransaction(signedTx.
rawTransaction);
  })
  .then(receipt => {
    console.log('Transaction receipt:', receipt);
  })
  .catch(error => {
    console.error('Error:', error);
  });
```

In this example, we used the account's private key to sign a transaction and then sent the signed transaction to the Ethereum network using `web3.eth.sendSignedTransaction()`.

Managing Ethereum accounts and wallets securely is essential to protect our funds and assets. Always follow best practices for private key management and avoid exposing private keys in public repositories or insecure environments. Use hardware wallets or secure key management solutions whenever possible to ensure the highest level of security for your Ethereum accounts.

In the next section, we'll expand the maintenance practices of an Ethereum network on GKE with more tools for troubleshooting potential issues that may arise during the execution of applications on the network.

Troubleshooting and maintaining the Ethereum network on GKE

Troubleshooting and maintaining an Ethereum blockchain network on GKE are critical aspects to ensure the smooth operation and reliability of the deployed network. The following guidelines, along with scripts when relevant, will help us troubleshoot issues and perform regular maintenance tasks.

The first set of guidelines focuses on monitoring the GKE cluster:

- **Monitoring and logging**

  Set up monitoring tools such as Prometheus and Grafana to track the health and performance of the GKE cluster and Ethereum nodes.

  Monitor metrics such as CPU usage, memory, network traffic, and blockchain-specific metrics such as pending transactions and block times.

- **Kubernetes dashboard**

  Use the Kubernetes dashboard to get insights into the state of your cluster, inspect pod logs, and diagnose issues.

- **Troubleshoot pods**

  Check the pods log files to troubleshoot issues with Ethereum nodes, and identify errors or abnormal behavior.

The following bash script opens the log files of a specific pod identified by name inside the GKE cluster:

```
# Get pod names in the namespace (replace NAMESPACE_NAME with your
actual namespace)
kubectl get pods -n NAMESPACE_NAME


# View logs of a specific pod (replace POD_NAME and NAMESPACE_NAME)
kubectl logs POD_NAME -n NAMESPACE_NAME
```

The second set of guidelines is about business continuity:

- **Backup and restore**

  Regularly back up the blockchain data and configuration to a secure location. This allows the network to recover to its latest saved status in case of data loss or catastrophic events.

- **Version compatibility**

  Ensure that the Ethereum client version in use is compatible with the network and smart contracts.

  Update the clients regularly to take advantage of bug fixes and new features.

- **Diagnosing network connectivity**

  Ensure the Ethereum nodes can communicate with each other.

  Verify that pods can reach other pods and that the Ethereum network port (usually 8545) is accessible from outside the cluster.

The following bash script will help diagnose potential network connectivity issues between pods by sending a `ping` command to the IP address of the pod to test:

```
# Diagnose network connectivity issues between pods (replace
NAMESPACE_NAME)
kubectl exec -it POD_NAME -n NAMESPACE_NAME -- ping OTHER_POD_IP
```

The third set of guidelines is about security practices:

- **Node maintenance**

  Periodically check the health of individual Ethereum nodes and consider replacing malfunctioning nodes.

  Use GKE's auto-scaling feature to automatically replace failed nodes.

- **Security audits**

  Conduct security audits of the Ethereum nodes and GKE cluster regularly to identify and fix vulnerabilities.

- **Rolling updates**

  When performing updates or maintenance on the Ethereum nodes, use rolling updates to ensure minimal disruption to the network.

The following bash script uses the `kubectl set image` command to update an existing deployment with a new image:

```
# Update a deployment with a new image (replace DEPLOYMENT_NAME and
IMAGE_NAME)
kubectl set image deployment/DEPLOYMENT_NAME DEPLOYMENT_NAME=IMAGE_
NAME:TAG
```

> **Testnet and staging environments**
>
> Always utilize testnet and staging environments to test updates and changes before deploying them to the production Ethereum network (mainnet).

Remember that these guidelines are not exhaustive, and troubleshooting and maintenance may vary based on the specific configuration and requirements of your Ethereum blockchain network on GKE. In my daily maintenance routine, I regularly review my network's performance, stay updated with the latest tools and practices, and take proactive measures to ensure the stability and reliability of my Ethereum network.

Summary

Hosting an Ethereum blockchain network on GKE offers developers a flexible and scalable environment for building decentralized applications. This chapter covered the key aspects of setting up, managing, and maintaining an Ethereum network on GKE to ensure a secure and reliable blockchain infrastructure.

Remember that running an Ethereum network on GKE is not much different than running any other solution. The GKE cluster still requires monitoring, securing, scaling, and all the best practices for an enterprise architecture designed for business continuity.

In the next chapter, we'll continue exploring the blockchain services in the Google Cloud, and we'll focus our attention on Blockchain Node Engine.

Further reading

- Google Kubernetes Engine documentation:

  ```
  https://cloud.google.com/kubernetes-engine/docs
  ```

- Geth documentation:

  ```
  https://geth.ethereum.org/docs
  ```

- Google Cloud Managed Service for Prometheus:

  ```
  https://cloud.google.com/stackdriver/docs/managed-prometheus
  ```

11
Getting Started with Blockchain Node Engine

In the ever-evolving landscape of technology, blockchain has gained widespread recognition among businesses as they seek innovative ways to leverage its potential for enhancing their operations. Google Cloud, a leading player in the cloud computing domain, has taken a significant step in this direction with the introduction of its **Blockchain Node Engine (BNE)**.

BNE in Google Cloud presents a cutting-edge solution for enterprises, developers, and blockchain enthusiasts looking to build, deploy, and manage blockchain networks efficiently and securely. By harnessing the power of Google's robust infrastructure and network, the engine offers a wide array of tools, services, and features to facilitate seamless blockchain deployment.

In this chapter, we'll delve into the fundamental concepts of BNE, its capabilities, and how it empowers businesses to harness the true potential of blockchain technology. We will explore its key components, including consensus algorithms, cryptographic security, and decentralized data distribution, shedding light on the essential attributes that make it stand out among other blockchain solutions.

Furthermore, we'll discuss the advantages of utilizing BNE over traditional methods of setting up and managing blockchain networks. From reduced operational complexities to enhanced scalability, we'll explore the various benefits that enterprises can expect by integrating their blockchain projects with this state-of-the-art engine.

Let's now dive into the following main topics:

- Introduction to BNE

- Features and benefits of BNE

- Using BNE for blockchain solutions

- Building a **decentralized app** (**dapp**) to interact with the blockchain node

- Integrating BNE with other GCP services

- Best practices for implementing blockchain solutions with BNE

Technical requirements

To run the scripts presented in this chapter, we need an account in Google Cloud. If you don't already have a Google Cloud account, sign up for one at `https://cloud.google.com/`. You may need to provide billing information, but Google Cloud offers a free trial with a credit for new users.

If not already available on your workstation, you also need to download the Google Cloud SDK from `https://cloud.google.com/sdk/docs/install` and install it as described in the online documentation.

All scripts presented in this chapter are available in the GitHub repository for the book: `https://github.com/PacktPublishing/Developing-Blockchain-Solutions-in-the-Cloud/tree/main/Chapter11`.

Introduction to BNE

BNE in Google Cloud stands at the forefront of cutting-edge distributed ledger technology, providing enterprises and developers with a powerful platform to deploy, manage, and scale blockchain networks with unparalleled efficiency and security.

At its core, BNE is designed to function as a **decentralized** computing environment, where multiple nodes collaborate to maintain the integrity and consensus of a blockchain network. By utilizing the robust infrastructure of Google Cloud, this engine offers a seamless and scalable solution to support a wide range of blockchain use cases, from financial applications to supply chain management and beyond.

One of the key features of BNE is its **flexibility** in supporting different types of blockchains and catering to different levels of transparency, decentralization, and access control that align with business objectives. However, it's also fair to say that BNE does not directly support consortium blockchain networks. This means it's not currently designed to work with private, permissioned blockchains used by a limited group of participants. BNE focuses on public blockchains, specifically Ethereum (currently **Proof-of-Stake** (**PoS**) Ethereum). This implies open participation and accessibility for anyone. There's no built-in functionality for consortium blockchains such as Hyperledger Fabric or

Quorum, commonly used for private networks. Some users have experimented with workarounds involving setting up a private network and connecting to it through BNE. This approach is complex and not officially supported by Google Cloud.

A critical aspect of any blockchain network is its **consensus algorithm**, which ensures that all nodes in the network agree on the validity of transactions and the state of the ledger. BNE supports various consensus mechanisms, including **Proof of Work (PoW)**, PoS, and **Practical Byzantine fault tolerance (PBFT)**, among others. This enables users to select the most suitable consensus algorithm for their specific use case, optimizing performance and resource utilization.

Security is paramount in any blockchain implementation, and BNE takes this aspect seriously. With Google Cloud's robust security infrastructure and cryptographic best practices, the engine ensures the integrity and confidentiality of data, safeguarding against potential attacks and unauthorized access.

Putting all together, the four key features of BNE, as depicted in the following figure, are decentralized architecture, the flexibility of a blockchain network support, security by design, and the adoption of multiple consensus algorithms to suit diverse business requirements:

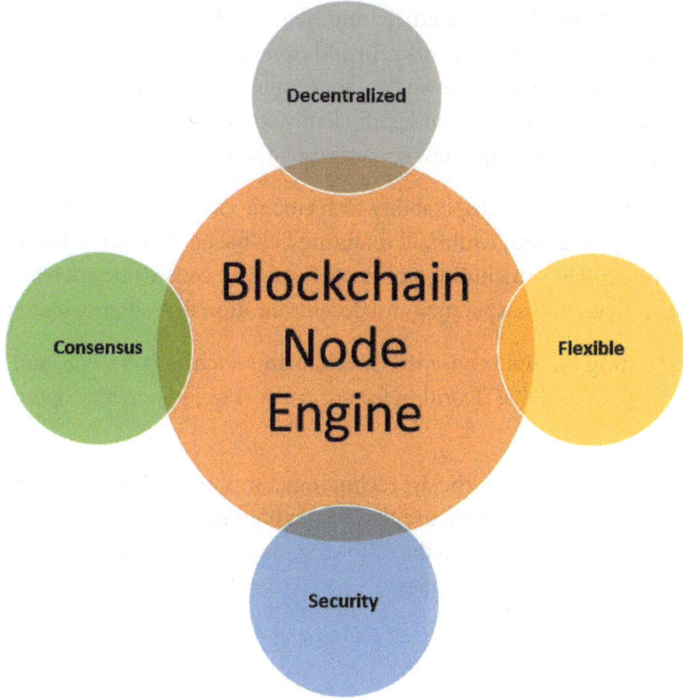

Figure 11.1 – Key features of BNE

The engine's management tools offer a user-friendly interface to monitor and control the blockchain network efficiently. Administrators can easily deploy new nodes, manage access controls, and perform updates seamlessly, streamlining the overall network management process.

Scalability is a crucial factor in blockchain adoption, especially as networks grow in size and complexity. BNE leverages Google Cloud's autoscaling capabilities to handle increasing workloads, ensuring optimal performance and responsiveness even during peak traffic times.

In addition to providing a feature-rich environment for blockchain network deployment, the engine also integrates with Google Cloud's extensive suite of services. This opens up opportunities for developers to leverage advanced data analytics, machine learning, and artificial intelligence capabilities, enhancing the value proposition of their blockchain applications.

In conclusion, BNE in Google Cloud empowers enterprises and developers to harness the transformative potential of blockchain technology without the burden of managing complex infrastructure. Its seamless integration with Google Cloud's powerful ecosystem, coupled with its flexibility, security, and scalability, makes it a compelling choice for businesses seeking to build innovative and efficient blockchain-based solutions.

Features and benefits of BNE

BNE in Google Cloud offers a host of features and benefits that make it a compelling solution for deploying and managing blockchain networks. In addition to the key features already identified in the previous section, BNE offers seamless network management. The engine provides user-friendly tools for deploying and managing blockchain nodes efficiently so that we can easily control access, monitor the network's health, and perform updates, streamlining the management process.

As we will see in the next section, interoperability with Google Cloud services allows developers to build robust and large-scale applications, while still retaining flexible blockchain support. The engine caters to various types of blockchains, including public, private, and consortium networks. This flexibility allows us to choose the level of transparency and decentralization that aligns with our specific use case.

In addition, by supporting multiple consensus mechanisms such as PoW, PoS, and PBFT, the engine enables us to select the most suitable algorithm for our network's requirements, optimizing performance and efficiency.

We understand that BNE is a state-of-the-art technology for the implementation of blockchain nodes in the cloud, but why exactly should we use this blockchain service over any other cloud-managed blockchain infrastructure? The benefits in the following pyramid will give us an idea of the operational advantages of BNE over similar technologies.

Figure 11.2 – Benefits of BNE

At the foundation, we have **reduced operational complexity**. BNE abstracts much of the underlying infrastructure, allowing businesses and developers to focus on their blockchain applications' core functionalities without getting bogged down in infrastructure management.

By utilizing the pay-as-you-go model of cloud computing, we can optimize its costs based on actual usage, avoiding the need for upfront investments in hardware and infrastructure. The engine's **robust security** measures and decentralized architecture contribute to building a high level of trust among participants in the blockchain network, ensuring the immutability and integrity of transaction records.

Google Cloud's extensive network of data centers worldwide enables BNE to provide **low-latency access and improved performance** to users across the globe. With all these advantages, BNE is an ideal platform to support various use cases of utilization of blockchain technology in the enterprise space. From supply chain management and financial applications to healthcare and identity verification, the engine is suitable for all of these types of applications. In the next section, we'll dive into the details of the implementation of such solutions.

Using BNE for blockchain solutions

Setting up BNE in Google Cloud involves a series of steps to ensure successful deployment. The following step-by-step guide will help us get started and configure BNE:

1. First of all, we need to sign into the Google Cloud at `https://console.cloud.google.com/`. Once we're in, let's search `Blockchain Node Engine` in the top product search bar. The **Blockchain Node Engine API** screen will display:

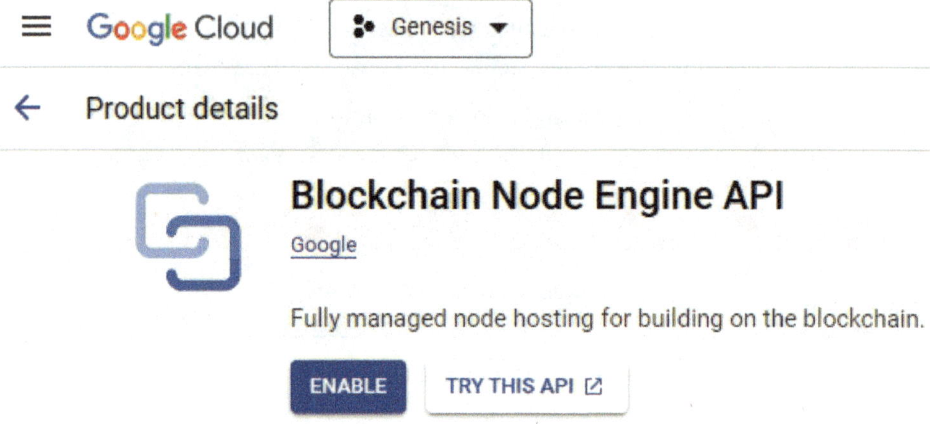

Figure 11.3 – The Blockchain Node Engine API product in Google Cloud

As we can read in the description, BNE is a **Fully managed node hosting for building on the blockchain**. The description is very vague, and it doesn't specify which blockchain is being used. However, by digging a bit more in detail in the official documentation available at `https://cloud.google.com/blockchain-node-engine/docs/`, we can find out that it's Ethereum, as expected. Thank you, Google!

> **Note**
>
> BNE has also introduced support for Polygon and Solana networks. At the time of writing, these two features are in preview and available to limited customers.
>
> Also, current support for blockchain is available in three GCP regions only: `us-central1`, `europe-west1`, and `asia-east1`.

2. Let's enable this service by selecting the **ENABLE** button. The next step that we're asked to complete is the creation of at least one node. We can create a blockchain node in two ways: either by carrying on using the console or by using an API call. Let's look at both options.

3. Using the Google Cloud console, we have to enter the required information in the screen depicted in the following picture, which opens when a new blockchain node is being created:

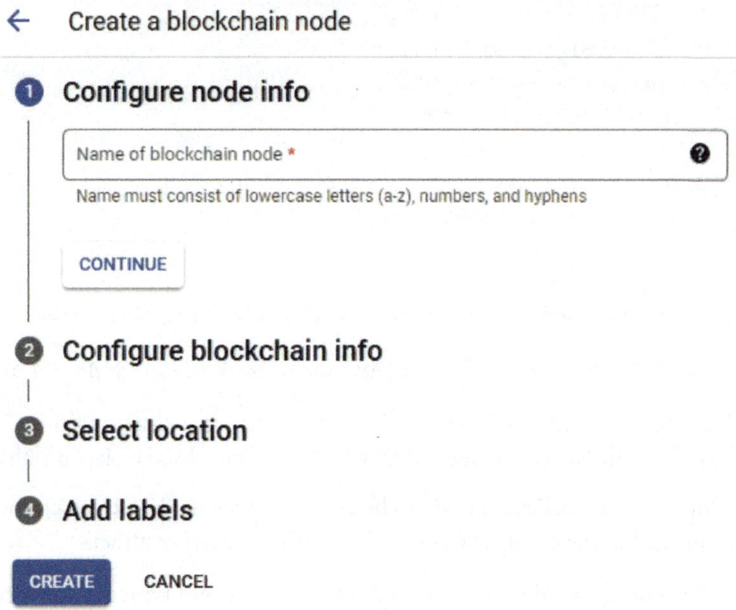

Figure 11.4 – Steps required to create a blockchain node

The functions of the four steps to create a blockchain node are as follows:

I.   *Step 1* is to assign a unique name to the node.

II. *Step 2* is for selecting the blockchain type (only Ethereum is available at this point unless you're a selected customer participating in the preview access program), the network to connect to (mainnet, Goerli Prater testnet, or Sepolia testnet), the node type (full node or archive), and the client to use (geth or Erigon).

III. *Step 3* is the selection of the deployment region. Currently, three regions are supported for this service: `us-central1` (Iowa), `asia-east1` (Taiwan), and `europe-west1` (Belgium).

IV. *Step 4*, which is optional, is to add key-value labels to the node being created.

4. Alternatively, we can also create a blockchain node by sending the following request:

```
curl -X POST \
  -H "Authorization: Bearer $(gcloud auth print-access-token)" \
  -H "Content-Type: application/json" \
  -d '{
    "blockchainType": "ETHEREUM",
    "ethereumDetails": {
      "consensusClient": " LIGHTHOUSE",
      "executionClient": "GETH",
```

```
      "apiEnableAdmin": false,
      "apiEnableDebug": false,
      "network": "MAINNET",
      "nodeType": "FULL"
    },
    "labels": {
      "LABEL_KEY": "LABEL_VALUE"
    }
  }' \
  https://blockchainnodeengine.googleapis.com/v1/projects/PROJECT_
  ID/\
  locations/LOCATION/blockchainNodes?blockchain_node_id=NODE_NAME
```

This API request creates a full node on the Ethereum `mainnet`, with a `geth` client. The name of the node is specified in the last line, in the `NODE_NAME` keyword. Please replace it with a suitable value.

It takes a few minutes for a blockchain node to be created. Once ready, we can connect to the node with the client indicated in the configuration, and deploy our smart contracts.

To connect to a BNE node and deploy a smart contract, we would typically use a combination of Web3 libraries and the blockchain node's API endpoint. Here is a simplified example using Web3.js, a popular JavaScript library for interacting with Ethereum blockchain. This example assumes that we have a smart contract ready to deploy.

First, if not already done, we need to install Web3.js using npm:

```
npm install web3
```

Now, let's assume we already have a simple smart contract written in Solidity, compiled with a tool such as Solc or Remix to get the **Application Binary Interface** (**ABI**) and bytecode needed for deployment.

We will use the Web3 library to connect to the BNE node over an HTTP or WebSocket endpoint provided by the Google Cloud console. In the following code snippet, replace `YOUR_BNE_NODE_ENDPOINT` with the actual BNE node endpoint:

```
const Web3 = require('web3');
const web3 = new Web3('YOUR_BNE_NODE_ENDPOINT');
```

To deploy the smart contract, we need the ABI and bytecode generated when compiling the Solidity contract. We also need an Ethereum account to deploy from, which requires having the account's private key. In the code snippet below, replace the `contractABI` and `contractBytecode` fields with the actual contract's ABI and bytecode, and `YOUR_ACCOUNT_PRIVATE_KEY` with the private key of the deploying account:

```
const contractABI = []; // Contract's ABI
const contractBytecode = '0x...'; // Contract's bytecode
const account = web3.eth.accounts
```

```
        .privateKeyToAccount('YOUR_ACCOUNT_PRIVATE_KEY');
web3.eth.accounts.wallet.add(account);
const deployerAddress = account.address;
const MyContract = new web3.eth.Contract(contractABI);
MyContract.deploy({ data: contractBytecode })
    .send({
        from: deployerAddress,
        gas: '1500000', // Adjust gas limit as necessary
        gasPrice: '30000000000', // Adjust gas price
    })
```

The following are very important to consider:

- **Security**: Be very careful with private keys. Never hardcode them in your scripts or expose them publicly.

- **Gas limit and gas price**: These values are set for demonstration purposes. Actual required gas and gas prices can vary depending on network congestion and contract complexity.

- **Testing**: Always test deployment on a test network (e.g., Rinkeby, Ropsten, etc.) before deploying on the main network.

Now that the blockchain part (node and contract) is ready, we can prepare a dapp with a UI, that interacts with the contracts deployed on the blockchain node. The next section describes how to build a dapp using Node.js on the backend, and the React framework for the frontend.

Building a dapp to interact with the blockchain node

This sample dapp demonstrates a basic interaction with the blockchain node just deployed in the previous steps. For this example, we'll use Node.js for the application backend and React.js for the frontend. The dapp implements the backend logic in the `app.js` file using Node.js and the Express (`https://expressjs.com/`) web framework, while the blockchain logic resides in the `blockchain.js` file. The frontend is created using React.js, which allows the user to add new blocks to the blockchain by entering data in the input field and clicking the **Add Block** button. The source code implementation is kept simple in the book for illustrative purposes. The full source code can be found in the GitHub repository for this chapter, available at `https://github.com/PacktPublishing/Developing-Blockchain-Solutions-in-the-Cloud/tree/main/Chapter11`.

Please save this JavaScript code in an `app.js` file. This script first obtains a list of existing blocks in the blockchain network by sending a `GET` request to the `blocks` endpoint exposed by the Express framework, and then it adds a new block by sending a `POST` request to the `addBlock` API endpoint:

```
const express = require('express');
const bodyParser = require('body-parser');
```

```
const Blockchain = require('./blockchain');

const app = express();
const blockchain = new Blockchain();

app.use(bodyParser.json());
app.get('/blocks', (req, res) => {
  res.json(blockchain.getChain());
});
app.post('/addBlock', (req, res) => {
  const data = req.body.data;
  blockchain.addBlock(data);
  res.json({ message: 'Block added successfully!' });
});
```

Now, let's create a `blockchain.js` file and save the following script. The script implements the actual blockchain logic in the `Blockchain` and the `Block` classes. This is obviously a simplified implementation of a block and its data to store on a blockchain, but it clarifies the kind of data required for each block, including a timestamp of when the block is created, the SHA-256 hash of its data, and the hash of the previous block in the blockchain:

```
const sha256 = require('sha256');

class Block {
  constructor(index, timestamp, data, previousHash) {
    // Initialize class variables
  }
}

class Blockchain {
  constructor() {
    this.chain = [this.createGenesisBlock()];
  }

  // Method prototypes only - full source code on GitHub
  createGenesisBlock()
  getLatestBlock()
  addBlock(data)
  getChain()
}
```

Lastly, the dapp is complete with its frontend UI built in React. We'll save the following script in the app-ui.js file. The axios library is used as a client for the exposed API. The UI presents a single **Add Block** button that, when selected, calls the handleAddBlock function. This function, in turn, makes a call to the addBlock API using the Axios client to add a new block to the blockchain:

```javascript
import React, { useState, useEffect } from 'react';
import axios from 'axios';

function SampledApp() {
  const [blocks, setBlocks] = useState([]);
  const [data, setData] = useState('');

  const handleAddBlock = async () => {
    await axios.post('/addBlock', { data });
    setData('');
    fetchBlocks();
  };

  return (
    <div>
      <input type="text" value={data} onChange={ (e) => setData(e.
target.value)} />
      <button onClick={handleAddBlock}>Add Block</button>
      <h2>Blocks</h2>
      <ul>
        {blocks.map((block) => (
          <li key={block.index}>
            Data: {block.data}, Hash: {block.hash}
          </li>
        ))}
      </ul>
    </div>
  );
}
```

As mentioned before, the full source code of this dapp is available in the GitHub repository of this book.

The backend and frontend of the app are now done. The user interaction is simple; it's based on pressing a single button from the user interface, but the potential is high. For example, communication with the blockchain engine can be automated on page load or on the verification of specific events that don't require user interaction. This could enable scenarios for process automation with no human intervention. On this foundation, the potential is great, and it can be expanded by integrating additional cloud services, as we'll see in the next section.

Integrating BNE with other GCP services

There cannot be a great dapp without leveraging multiple cloud services to create a richer experience that involves data storage, analytics, logging and monitoring, and potentially the use of AI technology. Google Cloud offers all these capabilities, and the possibility to integrate BNE with other Google Cloud services to enhance the functionality and capabilities of blockchain applications. Here are some examples of how we can integrate BNE with other GCP services:

- **Storage**: We can use Google Cloud Storage buckets to securely store documents, media, and other data related to the dapp.

- **Firestore**: We can integrate Google Cloud Firestore, a scalable and serverless NoSQL database, to store and manage additional application data related to the blockchain network. This can be useful for storing metadata, user profiles, or other application-specific information.

- **Data analytics**: Use Google Cloud's BigQuery service to analyze the data stored in your blockchain. BigQuery allows us to run fast and SQL-like queries on large datasets, delivering valuable insights from the transaction data recorded on the blockchain. We will talk about BigQuery in detail in *Chapter 12*.

- **Functions**: Leverage Google Cloud Functions to create serverless event-driven functions that can be triggered by blockchain events or external events. This enables the dapp to automate certain actions or processes in response to specific events on the blockchain.

- **Service bus**: Use Google Cloud Pub/Sub to enable communication and messaging between various components of the blockchain network. Pub/Sub facilitates real-time data streaming and event-driven architectures, enhancing the overall responsiveness of the application.

- **Monitoring and logging**: Integrate the dapp with Google Cloud's Monitoring and Logging services to track the health and performance of the blockchain nodes and network. This allows us to proactively identify and resolve issues before they impact the application's functionality.

- **Machine learning**: Utilize Google Cloud's AI and machine learning services, such as Google Cloud AI Platform and TensorFlow, to develop predictive models based on blockchain data. We can create models to detect anomalies, predict trends, or perform other data-driven tasks.

These are just a few examples of the many possibilities for integrating the dapp deployed on BNE with other GCP services. The seamless integration of Google Cloud services allows us to build sophisticated and powerful blockchain applications that leverage the strengths of both blockchain technology and GCP's extensive suite of tools and services.

Giving examples of integration of the previously built dapp with each cloud service is beyond the scope of this book. However, two services are particularly interesting to analyze more in detail as they are commonly used when building a dapp. These two services offer off-chain data storage capability, and they are Google Cloud Storage and Google Cloud Firestore.

Off-chain data storage

In this example, we'll demonstrate how to integrate BNE with Google Cloud Storage for off-chain data storage. Off-chain data storage refers to storing data that is not part of the blockchain itself but is associated with blockchain transactions or smart contract operations. This allows the dapp to store larger or more sensitive data externally while still retaining references to it on the blockchain.

For this example, we'll use Node.js and the `@google-cloud/storage` library to interact with Google Cloud Storage. Let's start with installing the necessary packages by running the following bash script:

```
npm install @google-cloud/storage
npm install express
```

Next, we need to create a Google Cloud Storage bucket. Creating a Google Cloud Storage bucket using a bash script involves using Google Cloud SDK's `gsutil` command-line tool, which is part of the Google Cloud SDK (see the *Technical requirements* section at the beginning of this chapter for details of how to install the SDK). When running the script, please note down the bucket name and ensure you have the necessary credentials to access it:

```
# Your Google Cloud Project ID
PROJECT_ID="my-gcp-project-id"
# The name of the bucket you want to create. Bucket names must be
globally unique.
BUCKET_NAME="my-unique-bucket-name"
# The location where you want to create your bucket. For example,  "US"
or "EU".
LOCATION="US"


# Create the bucket
gsutil mb -p $PROJECT_ID -c "STANDARD" -l $LOCATION gs://$BUCKET_NAME/
```

Here's what each command does:

- `gsutil mb`: This creates a new bucket.

- `-p $PROJECT_ID`: This specifies the project with which the bucket will be associated.

- `-c "STANDARD"`: This specifies the default storage class for the bucket. This could be `"STANDARD"`, `"NEARLINE"`, `"COLDLINE"`, or `"ARCHIVE"`.

- `-l $LOCATION`: This specifies the location (such as a region) for the bucket.

- `gs://$BUCKET_NAME/`: The name of the bucket you want to create.

Save this script in a file, for example, `create-bucket.sh`, give it executable permissions with `chmod +x create-bucket.sh`, and then run it from your computer. Before running the script, make sure to authenticate yourself by running `gcloud auth login`, and set your default project by running `gcloud config set project [PROJECT_ID]` if you haven't done so yet. Obviously, don't forget to replace PROJECT_ID and BUCKET_NAME in the script with your actual Google Cloud project ID and desired bucket name.

Now, you can execute the script by bash running:

```
./create-bucket.sh
```

This will create a new bucket in Google Cloud Storage with the indicated name.

Let's move now to the backend in Node.js. We'll save the following JavaScript code in an `app.js` file. The following is only a partial snippet of the entire source code, which you can find in the GitHub repository for this chapter:

```javascript
const express = require('express');
const bodyParser = require('body-parser');
const { Storage } = require('@google-cloud/storage');
const app = express();
const storage = new Storage();
const bucketName = 'your-bucket-name'

app.use(bodyParser.json());

app.post('/upload', async (req, res) => {
    const data = req.body.data;
    const fileId = req.body.fileId;
    const bufferData = Buffer.from(data);
    const file = storage.bucket(bucketName).file(fileId);
    await file.save(bufferData);
});

app.get('/fetch/:fileId', async (req, res) => {
    const fileId = req.params.fileId;
    const file = storage.bucket(bucketName).file(fileId);
    const [dataBuffer] = await file.download();
    res.json({ data: dataBuffer.toString() });
});
```

The script defines two methods:

- `post`: This is the endpoint to upload data to Google Cloud Storage. This method converts the data to a buffer before uploading it and then uploads the data to Google Cloud Storage with the `file.save` instruction.

- `get`: This is the endpoint to fetch data from Google Cloud Storage with an asynchronous call to `file.download`. The fetched data is then returned as a JSON message in the endpoint's response.

The last step of the dapp is the frontend, which in this case is plain HTML and JavaScript. We like to keep things simple! In the <body> section of the HTML file, add the following:

```
<h1>Blockchain Data Storage Example</h1>
<textarea id="data" rows="4" cols="50"></textarea>
<button onclick="uploadData()">Upload Data</button>
<div>
  <h2>Fetched Data</h2>
  <div id="fetchedData"></div>
</div>

<script>
async function uploadData() {
   const data = document.getElementById('data').value;
   const fileId = generateFileId();

   const response = await fetch('/upload', {
     method: 'POST',
     headers: {
       'Content-Type': 'application/json'
     },
     body: JSON.stringify({ data, fileId })
   });

   const result = await response.json();
   console.log(result);
}

async function fetchData(fileId) {
   const response = await fetch(`/fetch/${fileId}`);
   const result = await response.json();
```

```
        const fetchedDataDiv = document.getElementById('fetchedData');
        fetchedDataDiv.textContent = result.data;
    }
  </script>
```

In this example, the backend (Node.js with the Express framework) provides two endpoints: `/upload` to upload data to Google Cloud Storage and `/fetch/:fileId` to fetch data from the storage instance. The frontend (HTML and JavaScript) allows the user to enter data in a text area and upload it to the storage bucket. The user can also fetch data associated with a specific `fileId` from the bucket.

This example demonstrates a basic integration of BNE with Google Cloud Storage for off-chain data storage. In a real-world scenario, you would have additional logic to associate `fileId` with specific blockchain transactions or smart contract operations to maintain the reference to the data stored externally.

Integrate Cloud Firestore

Leveraging the previous example, we'll do the same but using Google Cloud Firestore for off-chain data storage. Cloud Firestore is a NoSQL document database that allows dapps to store and manage data in a flexible and scalable manner. We'll modify the previous example and use Google Cloud Firestore for data storage instead of Google Cloud Storage.

First of all, let's install the necessary packages by running the following bash commands:

```
npm install @google-cloud/firestore
npm install express
```

Next, we need to set up an instance of Google Cloud Firestore. Creating a Firestore database in Google Cloud Project can be automated using Google Cloud SDK's `gcloud` command-line tool. The steps involved are implemented in the following bash script:

```
PROJECT_ID="my-gcp-project-id"
DATABASE_MODE="NATIVE"
LOCATION="us-central1"


# Initialize Firestore
# gcloud firestore databases create --region=$LOCATION
--project=$PROJECT_ID
```

Before running the script, make sure that you've installed the Google Cloud SDK, as described in the *Technical requirements* section at the beginning of this chapter, and that you've authenticated your Google Cloud account by running `gcloud auth login`.

Save this script in a file, for example, `create-firestore.sh`, give it executable permissions with `chmod +x create-firestore.sh`, and then run it from your computer. This will initialize a new Firestore database in the specified Google Cloud Project. To execute the script, simply bash run it:

```
./create-firestore.sh
```

The Node.js backend code is very similar to what we have seen before for Google Cloud Storage, with the difference that in this example we're going to use Firestore instead:

```
const express = require('express');
const bodyParser = require('body-parser');
const { Firestore } = require('@google-cloud/firestore');
const app = express();
const db = new Firestore();


app.use(bodyParser.json());


app.post('/upload', async (req, res) => {
    const data = req.body.data;
    const fileId = req.body.fileId
    await db.collection('files').doc(fileId).set({ data });
});


app.get('/fetch/:fileId', async (req, res) => {
    const fileId = req.params.fileId;
    const doc = await db.collection('files').doc(fileId).get();
    res.json({ data: doc.data().data });
});
```

The two methods defined in the script are as follows:

- `post`: This is to upload data to Cloud Firestore, using the `db.collection('files').doc(fileId).set({ data })` function, where db is the Firestore object instance

- `get`: This is to fetch data from Cloud Firestore, using the `db.collection('files').doc(fileId).get()` function

In this example, the backend (Node.js with Express) uses `@google-cloud/firestore` to interact with Google Cloud Firestore. The endpoints for uploading and fetching data are similar to the previous example but are modified to use Firestore's database operations. There are no changes to the frontend HTML and JavaScript, as all the logic to handle Firestore over Storage is in the backend. How cool is that?

Best practices for implementing blockchain solutions with BNE

Implementing blockchain solutions with BNE in Google Cloud requires careful planning and adherence to best practices to ensure the security, scalability, and efficiency of your blockchain network. Best practices are not only about technical details. Compliance and change management also represent key aspects to consider in the successful implementation of a blockchain solution. The following figure depicts some of the best practices to consider, which I collected over the years and across multiple customers in a variety of industries:

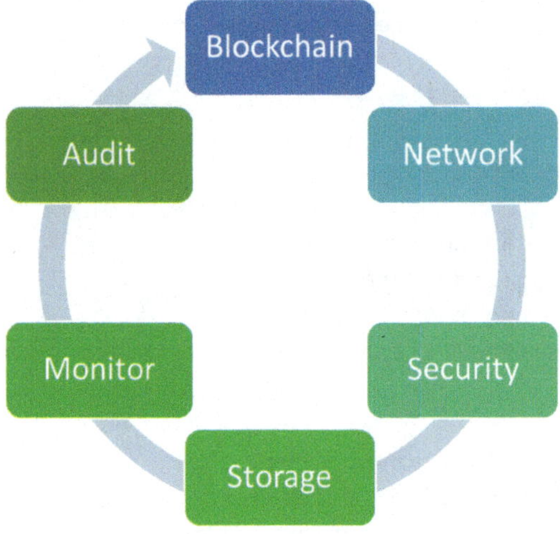

Figure 11.5 – Key best practices for blockchain solution with BNE

Overall, based on my experience, I would define the following functional and non-functional requirements when designing a blockchain solution:

- **Define clear objectives**: Clearly outline the objectives and use cases of your blockchain solution. Understand the specific problems you aim to solve and the value blockchain will add to your application.

- **Choose the right consensus algorithm**: Select the consensus algorithm that best suits your use case. Consider factors such as scalability, performance, and security when choosing between PoW, PoS, PBFT, or other consensus mechanisms.

- **Set up proper network permissions**: Implement robust identity and access management controls to ensure that only authorized participants can access and contribute to the blockchain network.

- **Use secure key management**: Protect your private keys using **Hardware Security Modules (HSMs)** or other secure key management systems. Proper key management is crucial for securing access to blockchain nodes.

- **Regularly update and patch nodes**: Keep your blockchain nodes up to date with the latest security patches and software updates to mitigate potential vulnerabilities.

- **Stay updated with blockchain technology**: Keep yourself and your team updated with the latest advancements and best practices in blockchain technology to continuously improve your implementation.

- **Monitor and audit**: Implement monitoring and logging to track the health, performance, and security of your blockchain nodes. Regularly audit your blockchain network to identify and address any issues promptly.

- **Monitor transaction fees**: Be aware of the transaction fees associated with your blockchain network and factor them into your cost calculations.

- **Optimize storage and data structure**: Optimize the data structure and storage mechanisms for efficiency and scalability. Consider using pruning techniques to remove old and unnecessary data from the blockchain.

- **Secure data off chain**: Utilize off-chain storage, such as Google Cloud Storage or Cloud Firestore, for large or sensitive data that doesn't need to be part of the immutable blockchain.

- **Plan for scalability**: Design your blockchain solution with scalability in mind. Use Google Cloud's autoscaling capabilities to handle increased workloads as your network grows.

- **Test and validate smart contracts**: Thoroughly test and validate your smart contracts to ensure they behave as expected and are free from vulnerabilities or bugs.

- **Implement disaster recovery and backup strategies**: Create disaster recovery and backup plans to protect against data loss and ensure the availability of your blockchain network.

- **Comply with regulations**: Ensure that your blockchain solution complies with relevant regulations and data privacy laws.

- **Educate participants**: Educate all participants about the functionalities, limitations, and security practices related to the blockchain network.

By following these best practices, you can build a robust, secure, and efficient blockchain solution with BNE in Google Cloud. Remember that each blockchain implementation is unique, so it's essential to adapt these best practices to suit your specific use case and requirements.

Summary

In this chapter, we discussed BNE in Google Cloud and its capabilities for deploying and managing blockchain networks. BNE offers a flexible and secure solution for various blockchain use cases, supporting different consensus algorithms and providing robust security measures. By leveraging Google Cloud's infrastructure and services, BNE simplifies the deployment and management of blockchain applications.

We also demonstrated how to integrate BNE with other Google Cloud services, such as Google Cloud Storage and Google Cloud Firestore for off-chain data storage. These integrations enable users to harness the full potential of GCP's offerings while leveraging the benefits of blockchain technology.

In the next chapter, we'll continue our analysis of additional services in Google Cloud by delving into the analytical capabilities of BigQuery, specifically in relation to blockchain data.

Further reading

- BNE documentation

  `https://cloud.google.com/blockchain-node-engine/docs/`

- Cloud Storage documentation

  `https://cloud.google.com/storage/docs`

- Firestore documentation

  `https://cloud.google.com/firestore/docs`

12

Analyzing On-Chain Data with BigQuery

As blockchain technology continues to evolve and permeate multiple sectors, the need to understand and leverage the vast amount of data these networks generate has never been more critical. The ability to analyze and extract meaningful insights from on-chain data can serve as a powerful tool for developers, researchers, and businesses alike, providing invaluable perspectives on the blockchain's operations, transactions, and user behaviors. But the question remains: how can we efficiently process and make sense of these enormous datasets?

This is where **Google Cloud Platform** (**GCP**), with its robust data analysis tool, BigQuery, comes into play. BigQuery allows users to examine on-chain data at scale, with the speed, power, and flexibility that the ever-evolving world of blockchain demands.

In this chapter, we'll delve into the exciting world of blockchain data analysis on GCP using BigQuery. We'll understand the complexities of on-chain data, explain the benefits of using BigQuery for data analysis, and provide you with practical examples of how to utilize this potent tool to uncover hidden trends, detect anomalies, and make data-driven decisions.

We'll dive into the following main topics:

- Introduction to BigQuery

- Features and benefits of BigQuery for on-chain data analysis

- Importing on-chain data into BigQuery

- Analyzing on-chain data with BigQuery

- Visualizing on-chain data with BigQuery

Technical requirements

To run the scripts presented in this chapter, you'll need a GCP account.

Visit the Google Cloud website at `https://cloud.google.com/` and sign up. You will need a Google account (for example, a Gmail account), and you will need to provide billing information. As a new user, you'll be given a free trial credit that you can use to explore GCP's services.

Once you've signed into the Google Cloud console, you need to create a new project. Click on the **Project** dropdown and select or create the project you will use to access BigQuery. You can create a new project by clicking on the **New Project** button at the top right of the dashboard. Provide a name for your project and click **Create**.

This book's GitHub repository (`https://github.com/PacktPublishing/Developing-Blockchain-Solutions-in-the-Cloud/tree/main/Chapter12`) contains the source code for all the examples in this chapter.

Introduction to BigQuery

BigQuery uses the power of GCP to deliver fast data warehouse capabilities at scale. The service is completely managed in the cloud, which means that its serverless model abstracts the underlying architecture and allows for seamless scalability, making it an ideal solution for handling and analyzing vast amounts of data, up to petabytes, in near real time.

At the core of BigQuery lies a distributed architecture that separates storage from computing resources. This design enables the computing power to be scaled up or down independently of data volume, allowing for cost-effective and flexible analysis of big data. The great advantage of all this is that it's fully managed in GCP: being serverless, BigQuery requires no database administration—it automatically manages cloud resources, providing a high level of reliability, redundancy, and data protection.

The following figure lists BigQuery's powerful features:

Figure 12.1 – BigQuery's key features

Let's take a closer look at each key feature of BigQuery:

- **Standard SQL**: BigQuery uses SQL, a popular, user-friendly querying language, making it accessible for data analysts and developers alike

- **High-speed analysis**: BigQuery's underlying Dremel technology allows for the execution of SQL queries on large datasets in seconds

- **Machine learning and AI integration**: BigQuery ML enables users to create and execute machine learning models using SQL, while integration with Google's AI Platform allows more sophisticated modeling and predictive capabilities

- **Geospatial data analysis**: BigQuery GIS provides geospatial capabilities, enabling analysis of geographic data

- **Data transfer and connectivity**: BigQuery provides data transfer services, streaming functionality for real-time insights, and seamless integration with popular data integration, business intelligence, and ETL tools

- **Security**: BigQuery adheres to Google's high-standard security model, providing encryption at rest and in transit, identity and access management, and more

By mastering the use of BigQuery, we can unlock valuable insights from our data in a fast, cost-effective, and scalable manner, regardless of whether we are handling gigabytes or petabytes of information.

Setting up BigQuery in GCP

Setting up BigQuery in GCP is a straightforward process. Let's walk through the step-by-step guide.

Enabling the BigQuery API

Before we can start using BigQuery, we need to enable the BigQuery API for our project:

1. Click on the navigation menu (hamburger icon in the top-left corner).

2. Hover over **APIs & Services** and click on **Library**:

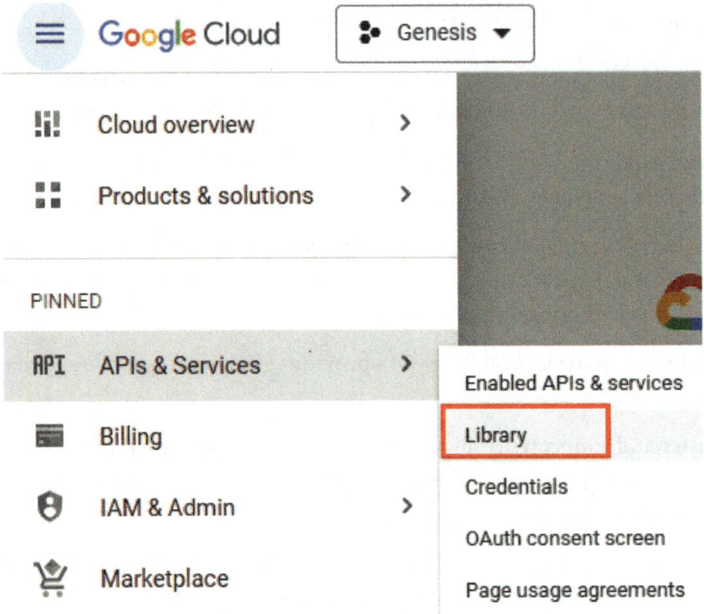

Figure 12.2 – APIs & Services > Library

3. In the API library, search for `BigQuery API` and click on it.

4. You will be taken to the **BigQuery API** page. Click on **MANAGE** to activate the BigQuery API for your project:

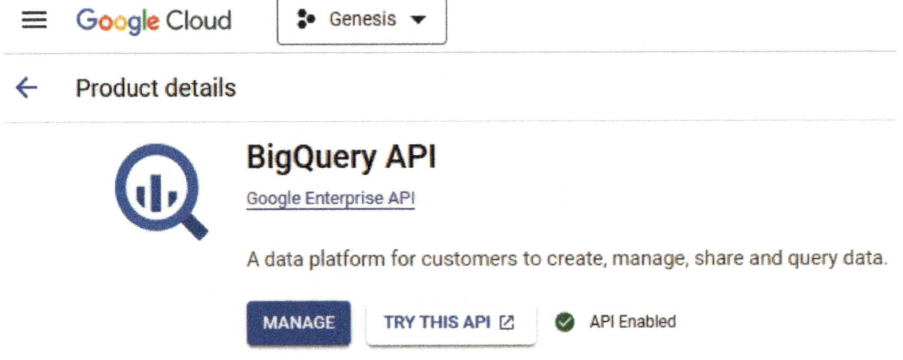

Figure 12.3 – The BigQuery API in Google Cloud

Opening the BigQuery SQL workspace

To access the BigQuery SQL workspace, we must open the navigation menu again and hover over the **BigQuery** menu item. **SQL workspace** is the first item on the list:

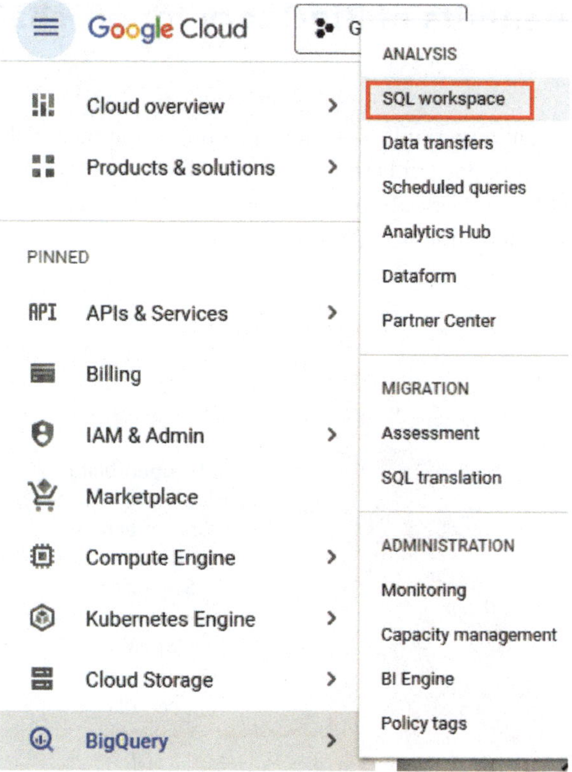

Figure 12.4 – Accessing BigQuery's SQL workspace

Now that we're in the SQL workspace, we can start composing and running SQL queries on existing datasets.

Creating a dataset

Before we can create a table and load data into BigQuery, we need to create a dataset:

1. In the left sidebar, click on the project's name.
2. Click on **Create Dataset**.
3. In the pop-up form, we need to give the dataset a unique ID and configure other settings, such as data location, default table expiration, and others.

Creating a table and loading data

We can now create a table within the dataset and load data into it. We can upload data from a local file, from **Google Cloud Storage (GCS)**, or even from another BigQuery table.

BigQuery also provides the capability to query data directly from external data sources, such as GCS, without the need to load the data into BigQuery first.

Features and benefits of BigQuery for on-chain data analysis

Analyzing on-chain data can pose significant challenges due to the sheer volume and complexity of data. Google BigQuery, with its powerful features and capabilities, can transform these challenges into opportunities. Let's explore the specific features and benefits of BigQuery when it's used for on-chain data analysis, as illustrated in the following figure:

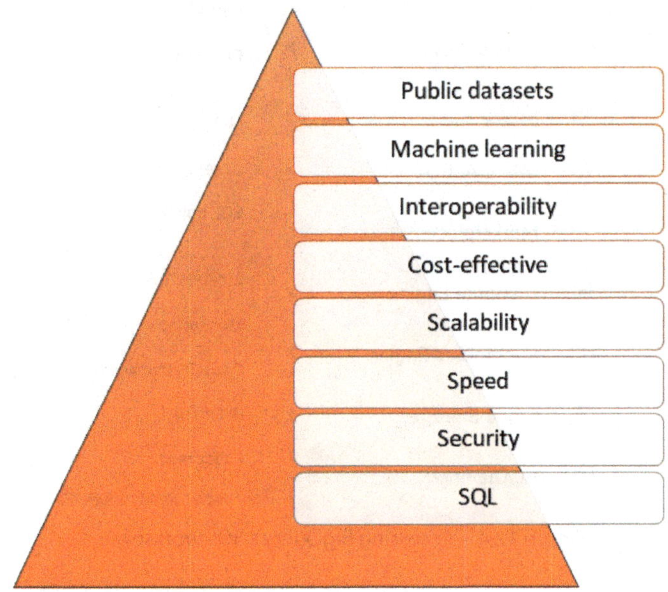

Figure 12.5 – Features and benefits of BigQuery for on-chain data analysis

Let's expand each feature and provide more details:

- **Public datasets**: Google Cloud maintains a program where public blockchain data is regularly updated in BigQuery format. This feature makes a wealth of on-chain data readily available for analysis.

- **Machine learning**: BigQuery's integration with Google's machine learning tools allows analysts to create predictive models directly from on-chain data. This can help with identifying trends, predicting future behavior, or detecting anomalies within the blockchain network.

- **Interoperability**: BigQuery provides seamless integration with popular data integration, visualization, and ETL tools, which makes it easier for users to ingest, analyze, and visualize blockchain data.

- **Cost-effective**: BigQuery's pay-as-you-go pricing model means you only pay for the queries you run, making it a cost-effective choice for on-chain data analysis.

- **Scalability**: Blockchain datasets can be extraordinarily large and continue to grow with each new block. BigQuery is designed to handle large-scale datasets, providing the ability to scale up or down as required, ensuring that even as the blockchain grows, your ability to analyze the data keeps pace.

- **Speed**: BigQuery's serverless model and underlying Dremel technology enable rapid data analysis. This allows for near real-time analytics, which is particularly valuable in the fast-moving world of blockchain.

- **Security**: Security is critical when dealing with sensitive on-chain data. BigQuery offers robust security features including encryption, identity and access management, and more.

- **SQL**: With SQL being the interface for interaction, BigQuery makes it easier for analysts and developers to work with blockchain data, which often has complex structures. By using SQL as a query language, the barrier to entry is drastically reduced.

In summary, BigQuery's features, which include scalability, speed, ease of use, machine learning integration, robust security, and cost-effectiveness, make it an excellent tool for analyzing on-chain blockchain data. By leveraging these features, organizations and researchers can gain valuable insights from their blockchain data, driving better decision-making and innovation.

> **Billing alert in BigQuery**
>
> Google BigQuery is a powerful, serverless data warehouse that enables super-fast SQL queries when using the processing power of Google's infrastructure. However, users often encounter surprises in their billing due to misunderstood aspects or common pitfalls. One of the primary issues is the way BigQuery charges for data storage and query processing. Users are charged for the amount of data scanned by each query, not the amount of data returned. This means that poorly optimized queries that scan large volumes of data can lead to high costs, even if the query returns a small result set. Another common mistake is not managing data storage effectively; stored data incurs charges, and excessive or unoptimized storage (such as keeping large amounts of duplicate data or not deleting temporary tables) can increase costs. Additionally, frequent use of the `SELECT *` command, which scans all columns in a table, instead of selecting only the necessary columns, can lead to unnecessary data processing and higher charges. Finally, not taking advantage of cost control measures such as setting budgets, monitoring alerts, or using the BigQuery cost estimator to predict query costs can lead to unexpected billing amounts. Being aware of and adjusting these aspects can significantly help in managing and optimizing BigQuery costs.

Importing on-chain data into BigQuery

Before diving into the import process, note that Google already maintains a regularly updated dataset for several popular blockchains, such as Bitcoin and Ethereum, as part of its public datasets program. We can query these datasets directly in BigQuery without needing to do any imports. I recommend that you always check if the blockchain data that you're interested in analyzing is already part of these public datasets.

Also, importing on-chain data into Google BigQuery involves two main GCP services: GCS and BigQuery itself. To ensure a smooth and error-free import process, the service account that's associated with both `storage.Client()` and `bigquery.Client()` instances must be configured with the appropriate permissions. Improper permission settings for these service accounts are a common source of errors for new users. Let's clarify what's needed:

- To import data from GCS, the service account that's used by `storage.Client()` needs access to read data from the buckets where the on-chain data is stored. The minimum required permission for this task is **Storage Object Viewer**.

- When dealing with BigQuery, the service account that's used by `bigquery.Client()` must have permissions that allow it to create and write to BigQuery datasets and tables, as well as run queries. The required permissions include **BigQuery Data Editor**, for creating tables and inserting data into BigQuery, and **BigQuery Job User**, for running load jobs.

If you need to import your on-chain data into BigQuery, the process typically involves obtaining the blockchain data (usually in the form of a `.csv` or `.json` file), uploading it to GCS, and then loading it into BigQuery. Let's look at a high-level Python script that uses the Google Cloud client libraries to do this.

Firstly, install the necessary Python libraries by running the following bash commands, if you haven't done so already:

```
pip install google-cloud-storage
pip install google-cloud-bigquery
```

Now, let's assume we've got our blockchain data in a `.csv` file and we've uploaded this file to a GCS bucket. We can use the following Python script to load that data into BigQuery:

```
from google.cloud import bigquery
from google.cloud import storage

storage_client = storage.Client()
bigquery_client = bigquery.Client()

bucket_name = 'your-bucket-name'
bucket = storage_client.get_bucket(bucket_name)

blob_name = 'your-data.csv'
blob = bucket.blob(blob_name)
```

```python
client = bigquery.Client()

dataset_id = "your_dataset"
table_id = "your_table"

job_config = bigquery.LoadJobConfig(
source_format=bigquery.SourceFormat.CSV, skip_leading_rows=1,
autodetect=True)

# Get the URI for the blob
uri = f"gs://{bucket_name}/{blob_name}"

load_job = client.load_table_from_uri(
uri, f"{dataset_id}.{table_id}", job_config=job_config
)

# Wait for the load job to complete.
load_job.result()

destination_table = client.get_table(f"{dataset_id}.{table_id}")
print("Loaded {} rows.".format(destination_table.num_rows))
```

This script will load our data from the `.csv` file in GCS into a BigQuery table, detecting data types automatically (`autodetect=True`). Make sure you replace `your-bucket-name`, `your-data.csv`, `your_dataset`, and `your_table` with your actual GCS bucket name, filename, BigQuery dataset ID, and table ID, respectively.

Keep in mind that this is a basic script, and the specific details can vary depending on the format of your data, the specific blockchain you're working with, and the data types of the on-chain data. Be sure to check the Google Cloud documentation for more details and options when loading your data.

Querying Ethereum

Let's assume we're working on the Ethereum network for now. As mentioned previously, Google provides an up-to-date dataset for the Ethereum blockchain as part of their public datasets program. The dataset is named `bigquery-public-data.crypto_ethereum`. It is updated daily, and we can query it just like any other BigQuery dataset.

The following figure shows the Ethereum architecture diagram, as implemented in GCP, to extract data from the Ethereum network daily, stage the data into GCS, and then load it into BigQuery tables:

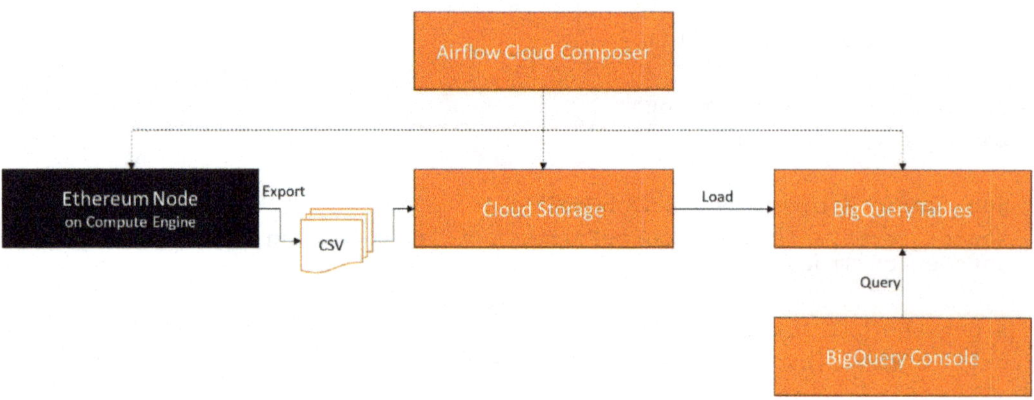

Figure 12.6 – Ethereum ETL architecture diagram

The following SQL query fetches the top 10 Ethereum addresses by the number of transactions:

```
SELECT
  `from`,
  COUNT(*) AS num_transactions
FROM
  `bigquery-public-data.crypto_ethereum.transactions`
GROUP BY
  `from`
ORDER BY
  num_transactions DESC
LIMIT 10
```

You can run this query directly in the BigQuery console by performing the following steps:

1. Go to the BigQuery console in your GCP dashboard.

2. In the left sidebar, click on the project name, then click on **Add Data | Explore public datasets**.

3. Search for Ethereum; you'll see the Ethereum Blockchain dataset.

4. Click **VIEW DATASET**. This will add the Ethereum dataset to your resources in the BigQuery console.

5. Now, you can write and run SQL queries directly on this dataset. Paste the preceding query into the **Query editor** area and click **Run**.

> **Cost control**
>
> Please remember that even though the data is public, you are still charged for the queries you run, so be mindful of your usage. For cost-effectiveness, design your queries so that they only retrieve the data that you need.

This dataset contains several tables, including the following:

- `blocks`: Data about each block

- `transactions`: Information about the transactions in each block

- `logs`: Event logs produced as the result of transactions

- `traces`: The state of Ethereum, representing the computation of contracts

The actual tables and their contents may be updated, so it's always a good idea to examine the schema before running queries. We can do this directly in the BigQuery console.

Querying Bitcoin

How about querying Bitcoin data? Google also provides a public dataset for Bitcoin blockchain data in BigQuery. The dataset is named `bigquery-public-data.crypto_bitcoin`. It's updated daily and can be queried just like any other BigQuery dataset:

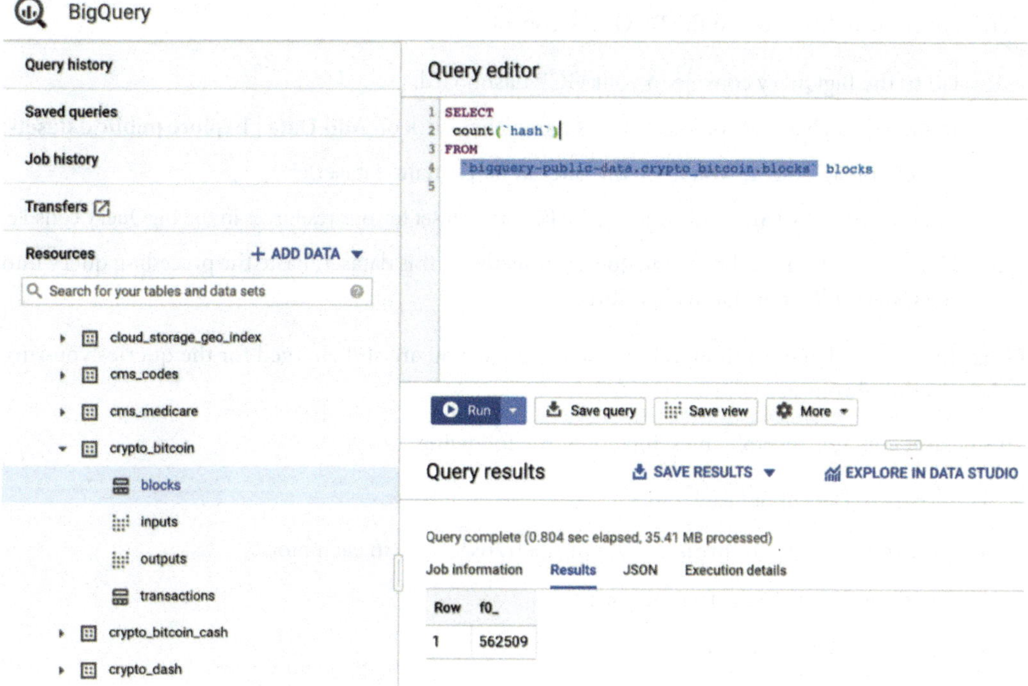

Figure 12.7 – Querying the Bitcoin public dataset in BigQuery

As seen for Ethereum previously, to query Bitcoin data, we can use a SQL query. The following query returns the top 10 Bitcoin addresses by the number of transactions:

```
SELECT
  output.output_pubkey_base58 AS address,
  COUNT(*) AS num_transactions
FROM
  `bigquery-public-data.crypto_bitcoin.transactions`,
  UNNEST(outputs) AS output
GROUP BY
  address
ORDER BY
  num_transactions DESC
LIMIT 10
```

This query counts the number of transactions for each address, ordering them in descending order to get the addresses with the most transactions.

Here's how to run this query in the BigQuery console:

1. Go to the BigQuery console in your GCP dashboard.

2. In the left sidebar, click on the project's name, then click on **Add Data | Explore public datasets**.

3. Search for `Bitcoin`; you'll see the Bitcoin blockchain dataset.

4. Click **VIEW DATASET**. This will add the Bitcoin dataset to your resources in the BigQuery console.

5. Now, we can write and run SQL queries directly on this dataset. Paste the preceding query into the **Query editor** area and click **Run**.

Please remember that even though the data is public, you are still charged for the queries you run, so be mindful of your usage.

This dataset contains several tables, including the following:

- `blocks`: Data about each block

- `transactions`: Information about the transactions in each block

- `inputs`: Inputs to the transactions

- `outputs`: Outputs of the transactions

The actual tables and their contents may be updated, so it's always a good idea to examine the schema before running queries. You can do this directly in the BigQuery console.

Analyzing on-chain data with BigQuery

Analyzing on-chain data with BigQuery provides a powerful way to extract insights from blockchain transactions, addresses, blocks, and more. Thanks to Google's public dataset program, data from popular blockchains such as Bitcoin and Ethereum are readily available for analysis, and the process can be streamlined using BigQuery's robust capabilities.

In the following examples, we'll identify a few use cases for the analysis of blockchain data and the relevant SQL queries for extracting such data in BigQuery.

The first step is to get familiar with the dataset's structure and understand the kind of information stored in each table. After that, we can run SQL queries to answer specific questions. For example, if we want to find the top 10 Ethereum addresses by the number of transactions, we could use a SQL query similar to the following:

```
SELECT
  `from`,
  COUNT(*) AS num_transactions
FROM
  `bigquery-public-data.crypto_ethereum.transactions`
GROUP BY
  `from`
ORDER BY
  num_transactions DESC
LIMIT 10
```

This SQL query counts the number of transactions by the `from` address in the `bigquery-public-data.crypto_ethereum.transactions` table and takes the first 10 rows. As you can see, this is a very standard SQL syntax that's common to most relational databases. BigQuery's SQL dialect is powerful and familiar to many users, making it easier to create complex queries.

Analyzing transaction patterns

You can use BigQuery to analyze transaction patterns. For instance, you might want to analyze the flow of cryptocurrency between addresses, detect patterns of fraudulent activity, or identify the most active periods for transactions. These analyses could help in developing trading strategies, improving network security, or forecasting network load.

Let's assume we want to analyze the Bitcoin transaction patterns to find out the average value of all outputs for each transaction over a specified date range. The following SQL query shows how we might do that:

```
SELECT
  t.block_timestamp AS date,
  COUNT(t.transaction_id) AS num_transactions,
```

```
    AVG(o.output_satoshis) / 100000000 AS avg_output_value_btc
FROM
    `bigquery-public-data.crypto_bitcoin.transactions` AS t,
    UNNEST(outputs) AS o
WHERE
    t.block_timestamp >= '2023-01-01 00:00:00 UTC'
    AND t.block_timestamp < '2023-02-01 00:00:00 UTC'
GROUP BY
    date
ORDER BY
    date
```

This query fetches the average value (in BTC) of all outputs for each transaction made in January 2023, as well as the total number of transactions for each day. The average value is computed by taking the average of the `output_satoshis` field (divided by 100,000,000 to convert Satoshi into BTC).

We can execute this query in the BigQuery console and then click **Run** to see the results.

Another point to keep in mind is that analyzing transaction patterns might involve a significant volume of data, especially for blockchains such as Bitcoin with a high number of transactions. This could lead to higher costs, so it's always a good idea to keep track of your query performance and costs in BigQuery.

On-chain analytics

BigQuery can also facilitate on-chain analytics, providing insights into blockchain operations. For instance, we could analyze the gas prices for Ethereum transactions over time, or how Bitcoin transaction fees evolve. This data can help users optimize their transactions and developers understand the network better.

Let's continue with our exploration of the Bitcoin blockchain data in BigQuery. A simple yet informative on-chain analysis might be to calculate the daily average transaction fee for a specified period. Transaction fees in Bitcoin are an essential part of the network as they incentivize miners to include transactions in the blocks they mine.

The following SQL query shows how we might calculate the daily average transaction fee in Bitcoin (in BTC) for January 2023:

```
SELECT
    block_timestamp_day,
    AVG((input.input_value - output.output_value)/100000000) as avg_tx_
fee_btc
FROM
    `bigquery-public-data.crypto_bitcoin.transactions` AS t
LEFT JOIN
```

```
   UNNEST(inputs) AS input
LEFT JOIN
   UNNEST(outputs) AS output
WHERE
   block_timestamp_day >= '2023-01-01' AND block_timestamp_day < '2023-
   02-01'
GROUP BY
   block_timestamp_day
ORDER BY
   block_timestamp_day
```

In this query, we calculate the transaction fee for each transaction by subtracting the total output value from the total input value (this is how transaction fees are calculated in Bitcoin). We then take the daily average of these fees. The resulting figure gives us an indication of the incentives being given to miners for this period.

These kinds of on-chain analytics can provide valuable insights into the operations of the blockchain network. By tracking changes in transaction fees over time, for example, you might be able to identify trends in network usage or miner behavior.

Please be aware that, similar to the previous example, depending on the range of dates you are analyzing, you may be querying a large amount of data, which may incur costs. Always consider this when running your analyses.

Predict the value of Bitcoin with machine learning

BigQuery's integration with Google Cloud's AI Platform means we can use machine learning models with our blockchain data. You might use this to predict future transaction volumes or identify anomalous transactions that could indicate fraudulent activity.

But how about predicting the value of Bitcoin using machine learning? Now, that sounds exciting! But I have to warn you: predicting the value of Bitcoin using machine learning models involves a multitude of factors and it is a complex problem. In practice, you should not only use on-chain data but also incorporate other variables, such as market sentiment, macroeconomic indicators, and so on. However, for the sake of simplicity, let's say we want to use the transaction fee data (that we calculated in the previous example) to predict the average transaction fee for the next day. We will use BigQuery's built-in linear regression function for this simple demonstration. Remember, this is a very simplistic model and may not give accurate predictions.

> **Warning**
>
> This example is a very simplistic model and may not give an accurate prediction. The author strongly discourages you from using the output of this example to make decisions concerning any form of cryptocurrency investment.

Here's the SQL code for this task:

```
CREATE OR REPLACE MODEL `your_project_id.dataset.model`
OPTIONS(model_type='linear_reg') AS
SELECT
  EXTRACT(DAYOFYEAR from block_timestamp_day) as day_of_year,
  AVG((input.input_value - output.output_value)/100000000) as avg_tx_
fee_btc
FROM
  `bigquery-public-data.crypto_bitcoin.transactions` AS t
LEFT JOIN
  UNNEST(inputs) AS input
LEFT JOIN
  UNNEST(outputs) AS output
WHERE
  block_timestamp_day BETWEEN '2023-01-01' AND '2023-03-01'
GROUP BY
  day_of_year
```

By executing this query, linear regression training is performed on the average daily transaction fee from January 1, 2023, to March 1, 2023. The model will try to predict `avg_tx_fee_btc` based on `day_of_year`.

Once the model has been trained, we can use it to make predictions using the following query:

```
SELECT
  day_of_year,
  predicted_avg_tx_fee_btc
FROM
  ML.PREDICT(MODEL `your_project_id.dataset.model`,
    (SELECT EXTRACT(DAYOFYEAR from block_timestamp_day) as day_of_year
      FROM `bigquery-public-data.crypto_bitcoin.transactions`
      WHERE block_timestamp_day BETWEEN '2023-03-02' AND '2023-03-31'
      GROUP BY day_of_year))
```

This will give us the predicted average daily transaction fee for each day in March 2023. Before running the query, don't forget to replace `your_project_id.dataset.model` with the actual project ID and dataset where you have stored your model.

Remember, the quality of predictions will heavily depend on the quality and relevance of input data, as well as the sophistication of your model. This example is a very basic one and may not yield accurate results. In a real-world scenario, we would likely need to use a more complex model and additional relevant features to get accurate predictions.

In summary, analyzing on-chain data with BigQuery involves understanding the dataset structure, writing SQL queries, and potentially using machine learning and data visualization tools. The combination of all these techniques allows us to extract meaningful insights from blockchain data, helping to drive strategy, optimize transactions, improve security, and much more.

Visualizing on-chain data with BigQuery

Finally, we can use data visualization tools such as Google Data Studio, Looker, or Tableau to create interactive dashboards with our BigQuery data. For example, we could create a real-time dashboard showing the most active Ethereum addresses, or how Bitcoin transactions are evolving.

Visualizing on-chain data with BigQuery can be achieved by integrating BigQuery with any of the visualization tools mentioned previously. For the sake of this example, we will use Google Data Studio due to its seamless integration with BigQuery and GCP as a whole.

Suppose we have the query from the earlier example that calculates the daily average transaction fee for Bitcoin transactions:

```
SELECT
  block_timestamp_day,
  AVG((input.input_value - output.output_value)/100000000) as avg_tx_
fee_btc
FROM
  `bigquery-public-data.crypto_bitcoin.transactions` AS t
LEFT JOIN
  UNNEST(inputs) AS input
LEFT JOIN
  UNNEST(outputs) AS output
WHERE
  block_timestamp_day >= '2023-01-01' AND block_timestamp_day < '2023-
02-01'
GROUP BY
  block_timestamp_day
ORDER BY
  block_timestamp_day
```

Now, we want to visualize these results. Here are the steps:

1. Access Google Data Studio.

    You can access Google Data Studio at `https://datastudio.google.com/`.

2. Create a new report.

    Click on the **Blank Report** option to start a new report.

3. Add BigQuery as a data source.

    Data Studio will ask you to add a data source to your report. Click on **BigQuery** and authorize Data Studio so that it can access your BigQuery data.

4. Configure the BigQuery data source.

    After choosing BigQuery, select your project, dataset, and view (this is the query result you saved as a view from your earlier work in BigQuery). After selecting these, click on the **Add** button at the bottom-right corner.

5. Visualize the data.

    After adding your data source, you'll be taken to a canvas where you can start creating your visualizations. The sidebar will show you a list of all the fields from your data source, and you can drag and drop these onto your canvas to create charts, tables, and other visualizations.

6. Display the time series chart.

    To plot the daily average transaction fee over time, you might use a time series chart. To do this, click on **Add a chart** and then select the **Time series** chart. Then, drag and drop the `block_timestamp_day` field into the **Dimension** field and `avg_tx_fee_btc` into the **Metric** field.

7. Customize and share.

    You can customize your visualizations, add more pages to your report, and, finally, share your report with others using the options in the top-right corner.

If you want to practice more with Data Studio and BigQuery data, there is a terrific tutorial by Google available at `https://codelabs.developers.google.com/codelabs/bigquery-data-studio`. This tutorial will guide you through the steps of visualizing BigQuery data in Data Studio, adding multiple visualizations, building a dashboard, and creating filters.

In conclusion, Google Data Studio provides a powerful tool for visualizing the results of on-chain data analyses from BigQuery. The ability to create interactive dashboards means you can create engaging, easy-to-understand presentations of your findings.

Summary

In this chapter, we discussed how to analyze on-chain data from blockchains using BigQuery, a serverless, highly scalable, and cost-effective cloud-based data warehouse offered by GCP.

After an introduction to BigQuery, where we detailed its architecture and features that are beneficial for on-chain data analysis, we went over the step-by-step process of setting up BigQuery in GCP. We then moved on to discussing how to import on-chain data into BigQuery, emphasizing that Google has made public datasets available for popular blockchains such as Bitcoin and Ethereum. We provided SQL query examples to help you access and analyze these datasets.

Finally, we discussed visualizing on-chain data with BigQuery. We illustrated how to integrate BigQuery with Google Data Studio to create dynamic and interactive visualizations, providing a step-by-step process of how to do it.

This chapter completes part four of this book. In the next chapter, we will start part five, which is dedicated to real-world use cases. Specifically, we will present a hands-on lab for building a decentralized marketplace on AWS.

Further reading

- *Ethereum in BigQuery: how we built this dataset*:

  ```
  https://cloud.google.com/blog/products/data-analytics/ethereum-
  bigquery-how-we-built-dataset
  ```

- *Ethereum in BigQuery: a Public Dataset for smart contract analytics*:

  ```
  https://cloud.google.com/blog/products/data-analytics/ethereum-
  bigquery-public-dataset-smart-contract-analytics
  ```

- *Bitcoin in BigQuery: blockchain analytics on public data*:

  ```
  https://cloud.google.com/blog/topics/public-datasets/bitcoin-
  in-bigquery-blockchain-analytics-on-public-data
  ```

- *Bitcoin Analytics using Google BigQuery*:

  ```
  https://bitcoindev.network/bitcoin-analytics-using-google-
  bigquery/
  ```

Part 5:
Exploring Real-World Use Cases and Best Practices

This part of the book will explore real-world use cases and best practices for implementing blockchain solutions on AWS, Azure, and GCP. It will provide hands-on labs and case studies that demonstrate the practical applications of cloud-native blockchain in various industries and use cases. A final chapter is dedicated to considerations of the future of blockchain in cloud applications, and the use of blockchain technology for good humanitarian purposes.

This part includes the following chapters:

- *Chapter 13, Building a Decentralized Marketplace on AWS*

- *Chapter 14, Developing a Decentralized Voting Application on Azure*

- *Chapter 15, Creating Verifiable Digital Ownership on GCP*

- *Chapter 16, The Future of Cloud-Native Blockchain*

13

Building a Decentralized Marketplace on AWS

Welcome to this hands-on lab, where you'll learn how to build a decentralized marketplace while leveraging the power and flexibility of **Amazon Web Services** (**AWS**) and blockchain technology. In today's digital age, centralization has often led to single points of failure, potential data breaches, and the overarching power of a single entity over a marketplace. This lab aims to introduce you to the technical foundations of creating a marketplace that is not only decentralized but also secure, transparent, and scalable.

In this chapter, we'll dive into the following main topics:

- The solution architecture for the decentralized marketplace

- The hosting infrastructure in AWS

- Setting up a blockchain network in AWS

- Creating and deploying the marketplace in AWS

Technical requirements

Building a blockchain solution often requires integration with third-party systems. In this case, we're going to leverage the following services in AWS, all of which complement the marketplace application by adding computing and storage capabilities:

- **Amazon Elastic Compute Cloud (EC2)**: Virtual servers in the cloud for running an Ethereum node

- **Amazon Relational Database Service (RDS)**: A relational database service for off-chain data

- **Amazon Simple Storage Service (S3)**: A storage service for unstructured data, such as files

We are going to run the marketplace on a blockchain network. Our choice for this lab is **Ethereum**. Ethereum is open source and can easily be deployed on **Virtual Machines (VMs)** in AWS. The relevant **smart contracts** will be published on the Ethereum *mainnet*, as described later in this chapter.

What to expect

This lab comprises step-by-step configuration guides, code samples, and best practices to ensure we successfully build and understand how to operationalize a decentralized marketplace. We'll be getting our hands dirty with code, configurations, and real-world scenarios, so be prepared for an interactive and enlightening experience.

This book's GitHub repository (`https://github.com/PacktPublishing/Developing-Blockchain-Solutions-in-the-Cloud`) contains the source code for the application presented in this chapter.

The duration of this lab is approximately 4-6 hours.

Prerequisites

For this chapter's lab, you'll require the following:

- A basic understanding of blockchain technology.

- Familiarity with AWS services.

- Experience in software development, preferably in languages such as JavaScript, Python, or Solidity.

- A working AWS account. If you don't have one, you can sign up for a free tier account here: `https://aws.amazon.com/`.

Solution architecture and hosting infrastructure

Building a decentralized marketplace on AWS involves several key aspects and software components that collectively contribute to the creation of a robust, scalable, and secure platform. The following figure depicts the key aspects of a decentralized app built on a public cloud service provider:

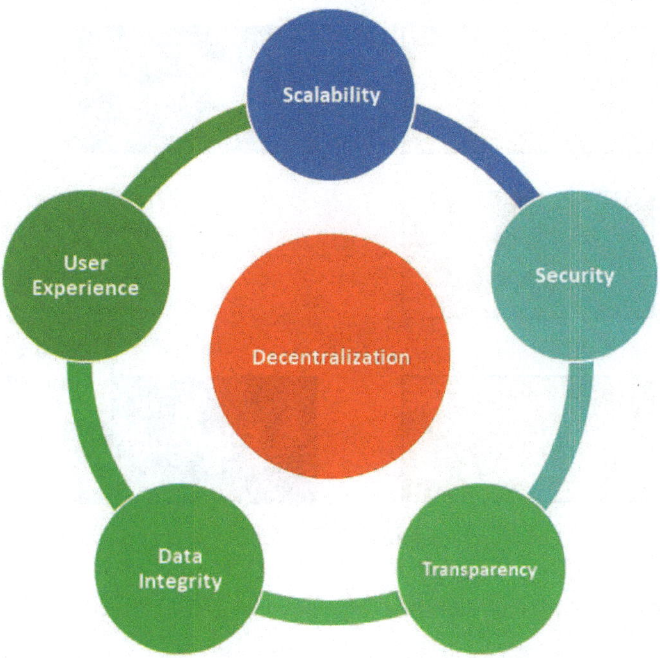

Figure 13.1 – Key aspects of a decentralized cloud-native app

We put **decentralization** at the core to provide the central architectural design decision to distribute control and management across a blockchain network, rather than having it centralized. Around the decentralized nature of the solution, the cloud service provider adds the following capabilities:

- **Scalability**: Utilizing AWS's cloud infrastructure allows the marketplace to scale resources up or down as needed

- **Security**: Using blockchain for transaction validation and AWS services for secure cloud operations

- **Transparency**: All transactions are recorded on the blockchain, ensuring full transparency within the marketplace

- **Data integrity**: Blockchain ensures that once the data is added, it can't be changed or removed, ensuring data integrity

- **User experience**: Despite the complex architecture, the user experience must remain intuitive and straightforward

Solution architecture

The solution we're building relies on the combination of traditional web components, such as frontend and backend services for managing user interaction and user data, and Web3 components such as Ethereum, smart contracts, and the MetaMask wallet. The following figure can be used as a reference for the designed solution:

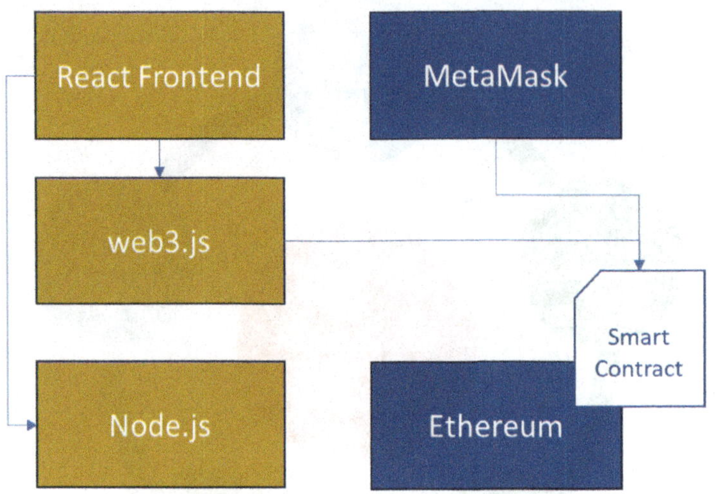

Figure 13.2 – Architecture reference for the designed solution

For the implementation of this marketplace solution, we have identified the following traditional web services:

- **Web interface**: A frontend developed using the React framework, which makes users interact with the marketplace's backend. React is purely a choice of convenience based on its popularity. You can use any other web frontend framework that you are familiar with.

- **web3.js**: A JavaScript library to connect the frontend with the Ethereum blockchain.

- **Node.js**: A backend service running on AWS EC2, serving as a middleman between the frontend and the various AWS and blockchain services.

On the web3.js side, the following services will be used:

- **Ethereum network**: The blockchain network where the marketplace's smart contracts will be deployed

- **Smart contracts**: Written in Solidity, these self-executing contracts manage the logic and state of the decentralized marketplace on the Ethereum blockchain

- **MetaMask**: A browser extension that allows users to make transactions that interact with the Ethereum blockchain easily

Hosting infrastructure on AWS

We are going to host the marketplace solution on AWS by leveraging some of the AWS services that we have already seen in *Chapters 4, 5,* and *6*.

Specifically, EC2 will host the Node.js API and potentially blockchain nodes if you choose to host your blockchain network. The frontend framework and backend services will communicate via API calls. These APIs are governed by **API Gateway**, which securely exposes the RESTful APIs for the frontend, at scale.

RDS will be used to manage user profiles, product metadata, and other off-chain information.

Finally, S3 will store unstructured data such as images, digital assets, or any other files related to the marketplace.

Optionally, **AWS Lambda** can be used for running serverless functions for specific automation tasks that run in the background, such as for sending notifications to users.

Security and monitoring

The marketplace solution can be completed with the following additional AWS services, which are meant to improve security and provide monitoring for the marketplace at runtime:

- **Identity and Access Management (IAM)**: To control access to AWS resources
- **CloudWatch**: For monitoring application and infrastructure performance
- **AWS WAF and Shield**: For additional security layers against web exploits

By combining these key aspects and software components effectively, we can build a decentralized marketplace that leverages the benefits of both blockchain technology and AWS services.

Setting up the blockchain network on AWS

Setting up the essential AWS services is crucial for the infrastructure of our decentralized marketplace.

Setting up Amazon EC2

Follow these steps to set up Amazon EC2 services:

1. Open the AWS management console at `https://aws.amazon.com/console/` and navigate to EC2 from the list of services. Click on **Launch instance** to create a new EC2 instance:

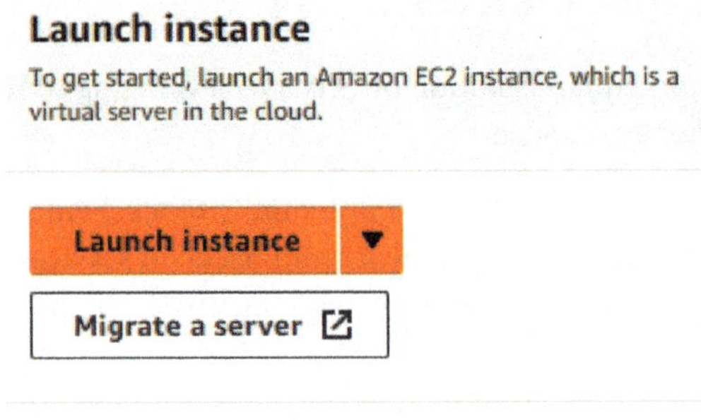

Figure 13.3 – Launching a new EC2 instance from the AWS management console

2. Select an **Amazon Machine Image (AMI)** according to your needs. For a basic setup, you can choose an Ubuntu server:

▼ **Application and OS Images (Amazon Machine Image)** Info

An AMI is a template that contains the software configuration (operating system, application server, and applications) required to launch your instance. Search or Browse for AMIs if you don't see what you are looking for below

Q *Search our full catalog including 1000s of application and OS images*

Quick Start

macOS	Ubuntu	Windows	Red Hat	SUSE Linux	De
Mac	ubuntu®	▦ Microsoft	🔴 RedHat	🦎 SUSE	d

Browse more AMIs

Including AMIs from AWS, Marketplace and the Community

Amazon Machine Image (AMI)

Ubuntu Server 22.04 LTS (HVM), SSD Volume Type   Free tier eligible
ami-0efcece6bed30fd98 (64-bit (x86)) / ami-03fd0aa14bd102718 (64-bit (Arm))
Virtualization: hvm   ENA enabled: true   Root device type: ebs

Description

Canonical, Ubuntu, 22.04 LTS, amd64 jammy image build on 2023-09-19

Architecture   AMI ID

64-bit (x86)   ▼   ami-0efcece6bed30fd98   Verified provider

Figure 13.4 – Selecting an Ubuntu server AMI

3. Choose an existing key pair, which is required to log into the VM, or create a new one and download it:

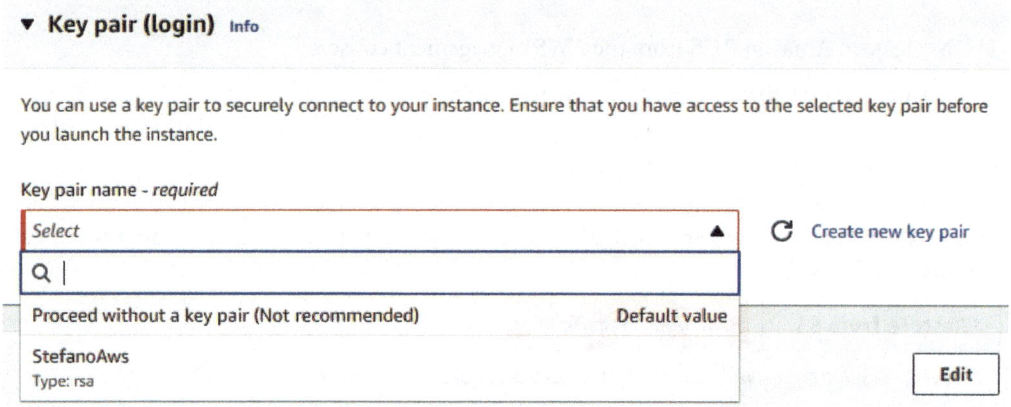

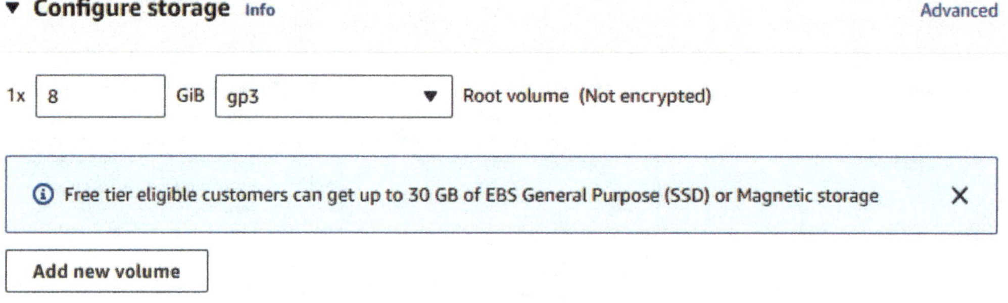

Figure 13.5 – Selecting or creating a key pair for logging in

4. Choose the instance type (for example, `t2.micro` for testing; this is more than enough for testing purposes). Add storage according to the specific needs. 8 GB of general-purpose SSD would work for a demo application. Applications in a live environment would need to be configured with more resources, depending on the expected volume of traffic and concurrent sessions:

Figure 13.6 – Adding storage to the selected AMI

5. Review the configuration in the **Summary** area and click **Launch instance**.

6. Select the instance you just created from the list of available EC2 instances and obtain its public IPv4 address. We'll need it later to access the VM via SSH, along with the key pair, by using the following command in the AWS Cloud Shell terminal:

```
ssh -i "Your-Key-Pair.pem" ubuntu@Your-EC2-IP-Address
```

Setting up Amazon RDS

Follow these steps to set up Amazon RDS services:

1. Navigate to Amazon RDS from the AWS management console.

2. Create a new database by selecting **Create database**:

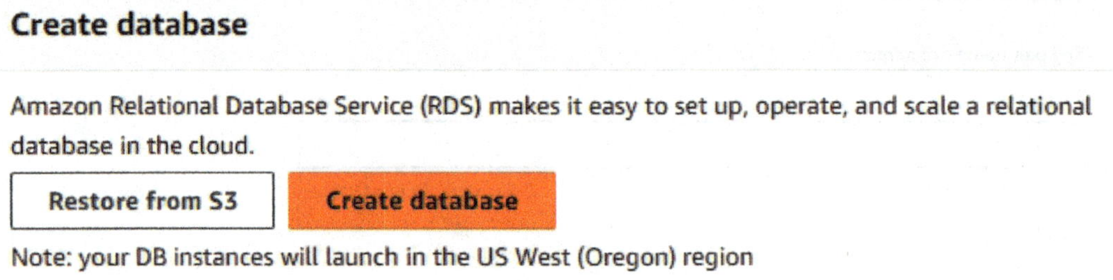

Create database

Amazon Relational Database Service (RDS) makes it easy to set up, operate, and scale a relational database in the cloud.

| Restore from S3 | **Create database** |

Note: your DB instances will launch in the US West (Oregon) region

Figure 13.7 – Creating a new RDS database

3. Choose a database engine (for example, MySQL) and pick one of the templates – that is, **Production**, **Dev/Test**, or **Free tier**:

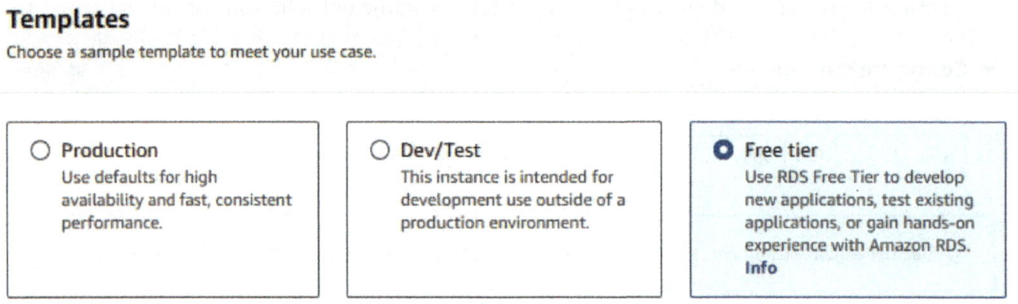

Templates

Choose a sample template to meet your use case.

○ Production	○ Dev/Test	● Free tier
Use defaults for high availability and fast, consistent performance.	This instance is intended for development use outside of a production environment.	Use RDS Free Tier to develop new applications, test existing applications, or gain hands-on experience with Amazon RDS. Info

Figure 13.8 – Choosing a template to meet your use case

4. Enter a unique name for the database instance and a login ID for the master user of the database instance. I recommend changing the proposed "admin" ID to something else. Optionally, we can use AWS Secrets Manager to safely store the user credentials. Otherwise, feel free to specify a strong master password:

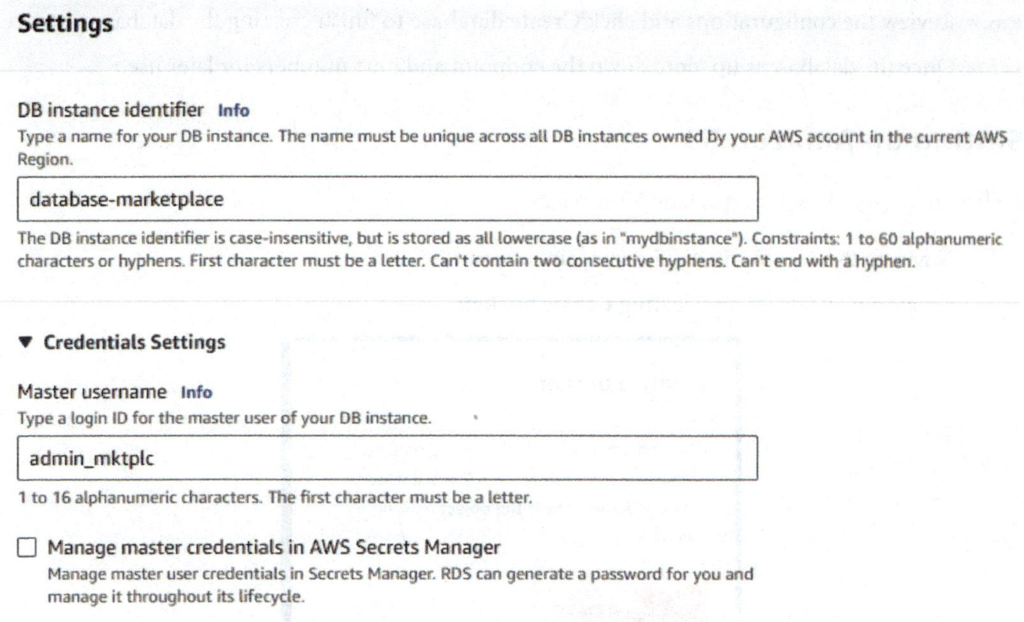

Settings

DB instance identifier   **Info**

Type a name for your DB instance. The name must be unique across all DB instances owned by your AWS account in the current AWS Region.

> database-marketplace

The DB instance identifier is case-insensitive, but is stored as all lowercase (as in "mydbinstance"). Constraints: 1 to 60 alphanumeric characters or hyphens. First character must be a letter. Can't contain two consecutive hyphens. Can't end with a hyphen.

▼ **Credentials Settings**

Master username   **Info**

Type a login ID for the master user of your DB instance.

> admin_mktplc

1 to 16 alphanumeric characters. The first character must be a letter.

☐ **Manage master credentials in AWS Secrets Manager**

Manage master user credentials in Secrets Manager. RDS can generate a password for you and manage it throughout its lifecycle.

Figure 13.9 – Entering a database name and master login credentials

5.   In the **Instance configuration** section, choose a database instance class from the available options. For MySQL on a free tier, I selected the **db.t3.micro** option with 2 vCPUs and 1 GiB of RAM. Storage is general-purpose SSD and 20 GiB of allocated space. Again, this sizing is perfect for testing and demoing purposes. It will need to be increased for live applications based on the performance and scalability requirements of the application:

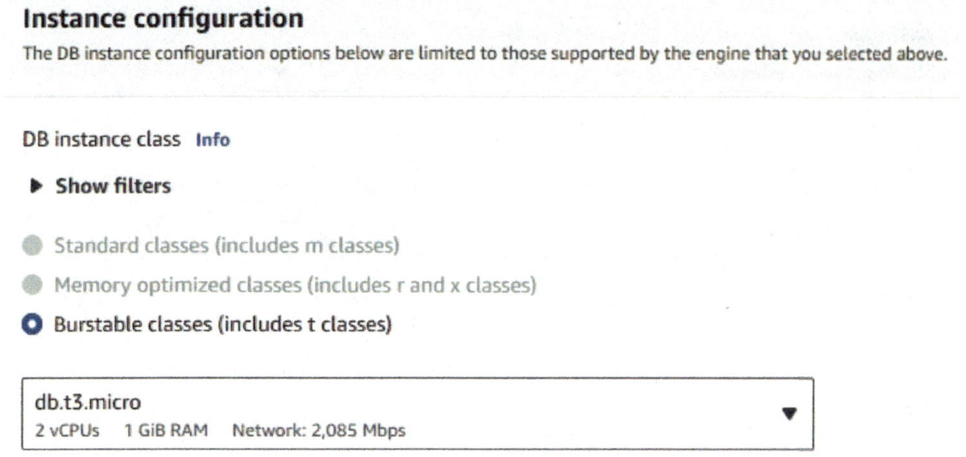

Instance configuration

The DB instance configuration options below are limited to those supported by the engine that you selected above.

DB instance class   **Info**

▶ **Show filters**

● Standard classes (includes m classes)

● Memory optimized classes (includes r and x classes)

🔘 Burstable classes (includes t classes)

> db.t3.micro
> 2 vCPUs   1 GiB RAM   Network: 2,085 Mbps

Figure 13.10 – Storage settings for the database instance

6. Review the configurations and click **Create database** to finish creating the database instance.

7. Once the database is up, note down the endpoint and port numbers for later use.

Setting up Amazon S3

Follow these steps to set up Amazon S3 services:

1. Navigate to S3 from the AWS management console.

2. Create a new bucket by selecting **Create bucket**:

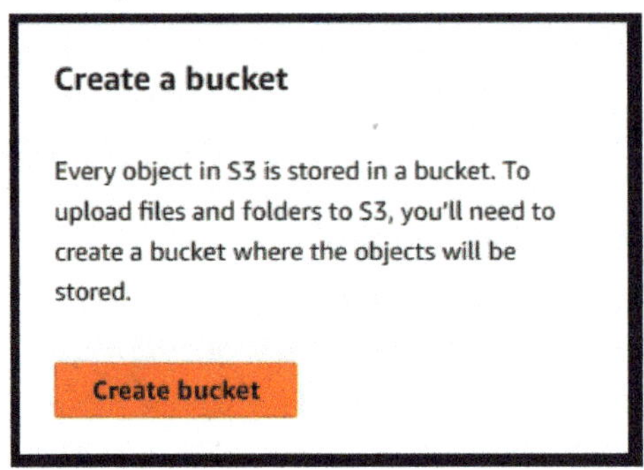

Figure 13.11 – Creating a new S3 bucket

3. On the **General configuration** section, choose a region where you wish to deploy the bucket and enter a unique name:

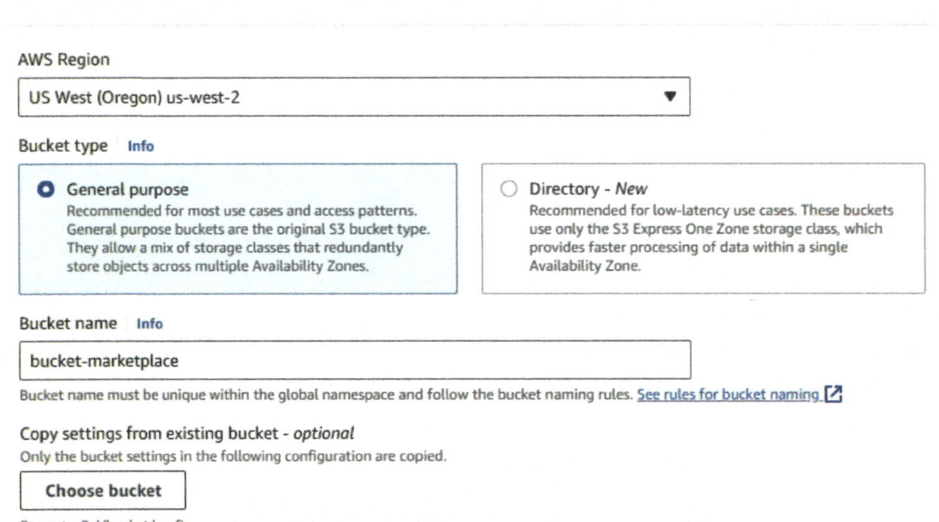

Figure 13.12 – General configuration of the S3 bucket

4. Configure permissions as per your needs. For **Object Ownership**, the recommended option is to have all the objects in the bucket owned by the current account, as opposed to access specified by using an **Access Control List (ACL)**. In addition, we want to disable all public access to objects in the bucket. The marketplace will authenticate into S3 to gain access to the bucket.

5. For this application, we will disable bucket versioning and select **Server-side encryption with Amazon S3 managed keys (SSE-S3)**:

Default encryption Info

Server-side encryption is automatically applied to new objects stored in this bucket.

Encryption type | Info

◉ Server-side encryption with Amazon S3 managed keys (SSE-S3)

○ Server-side encryption with AWS Key Management Service keys (SSE-KMS)

○ Dual-layer server-side encryption with AWS Key Management Service keys (DSSE-KMS)
Secure your objects with two separate layers of encryption. For details on pricing, see **DSSE-KMS pricing** on the **Storage** tab of the Amazon S3 pricing page. ⤴

Bucket Key

Using an S3 Bucket Key for SSE-KMS reduces encryption costs by lowering calls to AWS KMS. S3 Bucket Keys aren't supported for DSSE-KMS. Learn more ⤴

○ Disable

◉ Enable

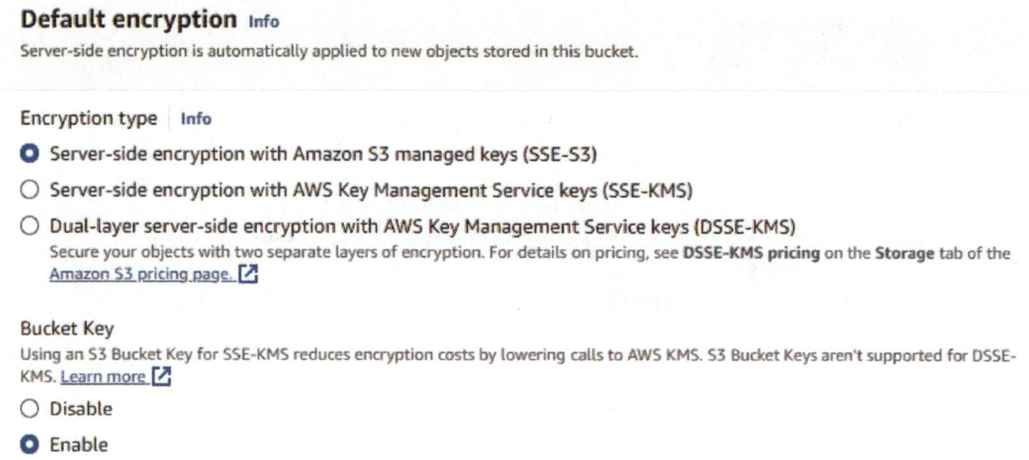

Figure 13.13 – Bucket versioning and default encryption settings

6. Review your configurations and click on **Create bucket** to finish creating the bucket.

We now have the essential AWS services configured. Next, we'll integrate these services into our decentralized marketplace, connect our EC2 instance to the RDS database, and utilize S3 for storing any off-chain data. By following these steps, we are well on our way to building a scalable and reliable decentralized marketplace.

Setting up Ethereum

Setting up a blockchain network on AWS involves several steps, including setting up an Ethereum node on an EC2 instance. This section provides a guide on how to do this, including some basic shell commands you can use to get started.

> **Disclaimer**
>
> This is a simplified example. Always remember to consider best practices for production-level deployments, such as securing the Ethereum node and using **Infrastructure as Code (IaC)** tools such as AWS CloudFormation or Terraform for resource provisioning.

Follow these steps to deploy an Ethereum node on AWS:

1. First, open a terminal and SSH into your EC2 instance by running the following bash command:

```
ssh -i "Your-Key-Pair.pem" ubuntu@Your-EC2-IP-Address
```

2. Update the Ubuntu package list and upgrade the system with the following bash commands. We need to have sudo-level permissions to be able to execute these commands:

```
sudo apt update
sudo apt upgrade -y
```

3. Install **Go Ethereum (Geth)**. Geth is the Go implementation of an Ethereum node. We can install it via the following bash commands (sudo-level permissions are required):

```
sudo apt install software-properties-common
sudo add-apt-repository -y ppa:ethereum/ethereum
sudo apt update
sudo apt install ethereum -y
```

4. Initialize the genesis block. First, create a directory where your blockchain data will be stored:

```
mkdir ~/my-ethereum-network
```

Then, create a `genesis.json` file with the following sample content:

```
{
  "config": {
    "chainId": 12345,
    "homesteadBlock": 0,
    "eip155Block": 0,
    "eip158Block": 0
  },
  "difficulty": "200",
  "gasLimit": "2100000",
  "alloc": {}
}
```

Initialize the genesis block with the following bash command:

```
geth --datadir ~/my-ethereum-network init ~/my-ethereum-network/
genesis.json
```

5. Next, run the Ethereum node using Geth:

```
geth --datadir ~/my-ethereum-network --networkid 12345
```

> **Note**
>
> This will start the Ethereum node in the foreground. We may want to run it as a background process or a service in a production environment.

6. Attach to the Ethereum node in a new terminal. We can interact with the node using the Geth console. Open a new terminal and SSH into the EC2 instance:

    ```
    geth attach ~/my-ethereum-network/geth.ipc
    ```

    We are now connected to our own private Ethereum blockchain network hosted on an AWS EC2 instance!

7. Optionally, enable the RPC API. We can also enable the HTTP-RPC server to interact with the node remotely or via web3.js by starting our Ethereum node with the following bash instructions:

    ```
    geth --datadir ~/my-ethereum-network --networkid 12345
    --http --http.addr '0.0.0.0' --http.port 8545 --http.api
    'eth,net,web3,personal'
    ```

> **Security warning**
>
> Be very careful when exposing the RPC API over the internet. Anyone can access your node; make sure you secure it adequately.

This section has helped us set up a private Ethereum blockchain on AWS EC2. The next steps will involve deploying our smart contracts to this network and building our decentralized marketplace on top of it.

Creating the decentralized marketplace application

Creating a frontend interface to interact with your deployed smart contract on the blockchain involves using web development frameworks and libraries that can communicate with Ethereum nodes. For this example, we'll use **React** and **web3.js**.

The prerequisites for executing the code in this section include having Node.js and npm installed, a smart contract deployed on the Ethereum node that we set up on AWS in the previous section, and MetaMask or another web3 wallet installed in our browser. Some basic familiarity with the React framework will also help us understand the code in detail.

The following steps will help us install the frontend of the marketplace application:

1. First, create a new React app by running the following bash command:

    ```
    npx create-react-app my-decentralized-marketplace
    ```

    Then, navigate to the project folder:

    ```
    cd my-decentralized-marketplace
    ```

2. Install the web3.js library so that you can interact with the Ethereum blockchain:

```
npm install web3
```

3. In the `src/App.js` JavaScript file, import `web3` and connect to the Ethereum network:

```
import Web3 from 'web3';


componentDidMount() {
  this.loadBlockchainData();
}


async loadBlockchainData() {
  if (window.ethereum) {
    window.web3 = new Web3(window.ethereum);
    await window.ethereum.enable();
  }


  // Load account
  const accounts = await window.web3.eth.getAccounts();
}
```

4. To interact with the smart contract, we need the **Application Binary Interface (ABI)** and the contract address. These can usually be found in the `build/contracts` directory of the Truffle project after compiling the smart contract's code:

```
const ABI = [...];  // Replace with your ABI array
const contractAddress = '0x...';  // Replace with your deployed
contract address


// Inside your loadBlockchainData() function
const web3 = window.web3;
const marketplaceContract = new web3.eth.Contract(ABI,
contractAddress);
```

5. Let's add methods to add a product and get a product from the smart contract:

```
async addProduct(name, price) {
  const accounts = await window.web3.eth.getAccounts();
  await this.state.marketplaceContract.methods.addProduct(name,
price).send({ from: accounts[0] });
}


async getProduct(index) {
  const product = await this.state.marketplaceContract.methods.
getProduct(index).call();
}
```

6. Now, we must add a simple UI to interact with the smart contract. This can go into the `render()` method inside the main React app component:

```
return (
  <div>
    <h1>My Decentralized Marketplace</h1>
    <button onClick={() => this.addProduct('Laptop', 1000)}>Add
Product</button>
    <button onClick={() => this.getProduct(0)}>Get Product</
button>
  </div>
);
```

This is a very basic example but should give us a decent starting point. You'd likely want to make this much more sophisticated with form validation, loading states, event subscriptions, and so forth.

The React app should now be set up to communicate with the smart contract on the Ethereum network. When we load the app, it should prompt us to connect to MetaMask. Once connected, we can add a product by clicking the **Add product** button and get a product by clicking the **Get product** button.

We now have a full-stack decentralized application: a smart contract deployed on a private Ethereum network running on AWS and a frontend built with React and web3.js, ready to interact with the smart contract.

Deploying the decentralized marketplace on AWS

The final step to complete this lab is to deploy the marketplace solution to the AWS services that we provisioned at the beginning. Deploying and especially automating the deployment of smart contracts can be challenging. For tools and best practices in deploying smart contracts, please refer to *Chapter 3* of this book.

Deploying a smart contract to a blockchain network is a key step in building a decentralized marketplace. The following instructions focus on the Ethereum blockchain, utilizing Solidity for the smart contract and Truffle as the development environment:

1. Install Truffle globally using npm:

```
npm install -g truffle
```

2. Create a new directory for the project and navigate to it:

```
mkdir MyDecentralizedMarketplace
cd MyDecentralizedMarketplace
```

3. Initialize the Truffle project:

```
truffle init
```

4. Create a new Solidity file under the `contracts` directory in the Truffle project:

    ```
    touch contracts/Marketplace.sol
    ```

    Edit `Marketplace.sol` and write the smart contract:

    ```solidity
    // SPDX-License-Identifier: MIT
    pragma solidity ^0.8.0;

    contract Marketplace {
        struct Product {
            string name;
            uint256 price;
        }

        Product[] public products;

        function addProduct(string memory _name, uint256 _price)
    public {
            Product memory newProduct = Product(_name, _price);
            products.push(newProduct);
        }

        function getProduct(uint256 _index) public view
    returns(string memory, uint256) {
            Product memory product = products[_index];
            return (product.name, product.price);
        }
    }
    ```

5. Compile the smart contract using Truffle:

    ```
    truffle compile
    ```

6. Edit the `truffle-config.js` file by specifying the network where the contract will be deployed. Here's an example of how you can set up the configuration for deploying to our Ethereum node running on AWS:

    ```javascript
    module.exports = {
      networks: {
        development: {
          host: "Your-AWS-EC2-IP",
          port: 8545,
          network_id: "12345", // Match the network id you used
    during Ethereum node setup
          gas: 3000000
        },
    ```

```
    },
    compilers: {
      solc: {
        version: "0.8.0",
      },
    },
  };
```

7. Before deploying, create a migration script in the `migrations` directory:

    ```
    touch migrations/2_deploy_contract.js
    ```

    Edit the file and add the following code:

    ```
    const Marketplace = artifacts.require("Marketplace");


    module.exports = function(deployer) {
      deployer.deploy(Marketplace);
    };
    ```

    Run the migration to deploy the contract:

    ```
    truffle migrate --network development
    ```

8. We can now interact with our deployed smart contract using Truffle's console and verify that the connection works:

    ```
    truffle console --network development
    ```

    Then, we can interact with the contract:

    ```
    let instance = await Marketplace.deployed()
    await instance.addProduct("Laptop", 1000)
    let product = await instance.getProduct(0)
    ```

Our smart contract is now deployed on our private Ethereum network running on AWS, and we have verified that we can interact with it. Along with the frontend UI presented in the previous section, this completes the lab for developing a decentralized marketplace. The source code for this solution can be found in this book's GitHub repository: `https://github.com/PacktPublishing/Developing-Blockchain-Solutions-in-the-Cloud`.

Summary

In this hands-on lab, we explored the key components of building and scaling a decentralized marketplace using AWS and blockchain technology. The lab started with setting up essential AWS services, including EC2 for hosting blockchain nodes, RDS for off-chain data storage, and S3 for storing large files. We then dived into deploying a smart contract on an Ethereum blockchain network.

Next, we developed a frontend interface using React.js and web3.js to interact with the deployed smart contract. The frontend allows for the addition and retrieval of products in the decentralized marketplace.

Ready for another hands-on lab? In the next chapter, we'll develop a decentralized voting application on Azure.

Further reading

- Amazon EC2 documentation:

  `https://docs.aws.amazon.com/ec2/`

- Amazon RDS documentation:

  `https://docs.aws.amazon.com/rds/`

- Amazon S3 documentation:

  `https://docs.aws.amazon.com/s3/`

14

Developing a Decentralized Voting Application on Azure

Welcome to this hands-on lab where we will explore the fascinating intersection of blockchain technology and cloud computing, focusing on Microsoft's Azure platform. As the demand for secure, transparent, and efficient systems continues to grow, decentralized technologies such as blockchain have stepped into the limelight. This lab is designed to give you practical experience in creating a decentralized voting application, providing a unique blend of cryptographic security measures offered by blockchain and the powerful computing resources available on Azure.

Why decentralized voting? Traditional voting systems, although generally reliable, are often subject to scrutiny regarding their transparency, security, and the possibility of fraud. Decentralized voting applications aim to mitigate these concerns by creating an immutable and transparent record of votes, accessible to anyone who wishes to verify the results, yet secure enough to ensure the privacy and integrity of each vote.

What will you learn? By the end of this lab, you will have a functional decentralized voting application running on Azure, and you'll have gained the skills needed to further explore the possibilities that blockchain technology can offer in various domains.

The lab is structured as follows:

- Introduction to developing a decentralized voting application on Azure

- Setting up the blockchain network on Azure

- Developing the voting application

Let's get started!

Technical requirements

To run the code and scripts presented in this chapter, you will need an account in the Azure cloud. A free account, which you can obtain at `https://azure.microsoft.com/en-us/free/`, will suffice.

All the source code is available in this book's GitHub repository: `https://github.com/PacktPublishing/Developing-Blockchain-Solutions-in-the-Cloud`.

Introduction to developing a decentralized voting application on Azure

Ensuring data integrity, voter privacy, and system robustness is essential for any voting application, particularly in a decentralized environment. Before we cover the implementation details of the proposed solution, I would like to illustrate a few common cryptographic techniques that can be used to improve data integrity, voter privacy, and overall system robustness. We can then implement some or all of these techniques in the smart contract of the voting application:

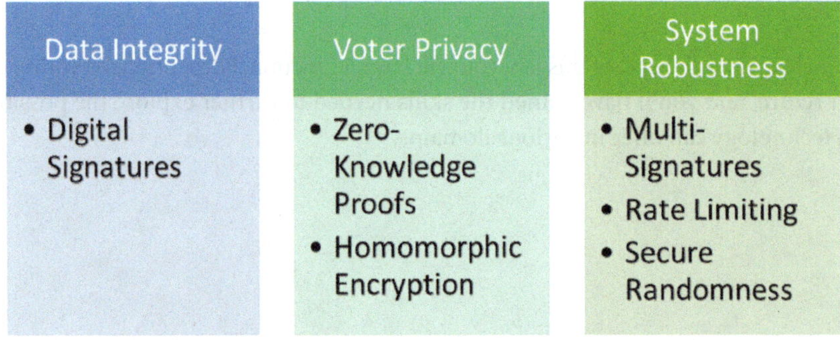

Figure 14.1 – Cryptographic techniques in a decentralized application

Data integrity

A very common approach to ensure data integrity in a decentralized application is to digitally sign and verify a transaction. **Digital signatures** are based on asymmetric encryption algorithms and the use of private and public keys.

In our voting app, every vote can be digitally signed using the voter's private key. This ensures that the vote hasn't been tampered with after being cast. A smart contract can then validate the signature using the public key before accepting the vote.

The following Solidity smart contract example validates a digital signature:

```
function castVote(bytes memory signature, uint256 candidateId) public
{
    address signer = recoverSigner(signature, candidateId);
    require(signer == msg.sender, "Invalid signature");
    // Continue with vote casting logic
}
```

Voter privacy

To enforce privacy in the vote, which otherwise would be openly visible on the blockchain, we can introduce one of two privacy-enforcing techniques:

- **Zero-Knowledge Proofs (ZKPs)**: ZKPs can prove that a transaction (such as a vote) is valid, without revealing the transaction's details. Technologies such as zk-SNARKs can be incorporated into smart contracts to add another layer of privacy.

  The following conceptual example of a Solidity smart contract illustrates how to validate a zk-SNARK proof without revealing voter details in a `castVoteWithZkSnark` function:

  ```
  function castVoteWithZkSnark(bytes memory zkProof) public {
      bool isValid = validateZkSnarkProof(zkProof);
      require(isValid, "Invalid zk-SNARK proof");
      // Continue with vote casting logic
  }
  ```

- **Homomorphic encryption**: Votes are encrypted in such a way that they can be tallied without being decrypted. This ensures voter privacy and allows for public verification of the tally.

  The smart contract can be extended with a `castVoteWithHomomorphicEncryption` function to store the encrypted vote, which will be decrypted only for tallying:

  ```
  function castVoteWithHomomorphicEncryption(bytes memory
  encryptedVote) public {
  }
  ```

Ethical implications and challenges of voter privacy in decentralized voting systems

While decentralized voting systems offer unprecedented levels of security and transparency, they also introduce complex ethical implications and challenges, particularly concerning voter privacy. The balance between ensuring a transparent electoral process and protecting individual privacy rights is delicate and requires careful consideration.

One of the core ethical challenges in decentralized voting systems is the tension between maintaining voter anonymity and ensuring the transparency of the voting process. Blockchain technology makes every transaction traceable and permanent, raising concerns about potential privacy breaches. While technologies such as ZKPs offer solutions by allowing voters to verify their vote without revealing their identity, ensuring these technologies are foolproof and accessible to all voters is a significant challenge.

Decentralized systems, by nature, distribute data across multiple nodes, potentially reducing the risk of mass surveillance and data collection by central authorities. However, the immutable nature of blockchain raises concerns about the permanent storage of sensitive information, even in encrypted forms. There's a risk that future advancements in technology could decrypt today's secure data, posing long-term privacy risks.

The ethical challenge of equity and accessibility in decentralized voting systems also ties into privacy concerns. Ensuring that all eligible voters have equal access to the voting system, regardless of their technological literacy or access to digital infrastructure, is crucial. Otherwise, the system could exclude significant portions of the population, undermining the democratic process and potentially skewing election results.

In conclusion, the ethical implications and challenges of voter privacy in decentralized voting systems highlight the need for a nuanced approach to technology implementation. Ensuring voter privacy while maintaining transparency and integrity requires continuous technological innovation, legal and regulatory frameworks that respect individual rights, and mechanisms to address potential abuses. As decentralized voting systems evolve, addressing these ethical challenges will be critical in realizing their potential to enhance democratic processes.

System robustness

Making sure that the system is robust and can respond safely to cyber attacks includes introducing a few techniques:

- **Multi-signatures for admin functions**: Stronger system robustness can be obtained by requiring multiple parties (instead of a single entity) to approve changes to the voting system, such as adding or removing candidates or ending the election early. This ensures that no single entity can control the election.

Our smart contract can accept multiple candidates in an `addCandidate` function that accepts the candidate's name as input and adds it to an internal list:

```
function addCandidate(string memory name) public
onlyMultiSigners {
}
```

- **Rate limiting**: Implement rate limiting on contract methods that are sensitive to abuse could prevent spam attacks.

  For example, the `castVote` function checks whether the maximum amount of votes that can be cast has been reached by a voting member:

```
function castVote(uint256 candidateId) public {
    require(voteCount[msg.sender] < MAX_VOTES, "Rate limit
exceeded");
    // Continue with vote casting logic
}
```

- **Secure randomness**: Using cryptographic techniques to generate secure randomness for any lottery-based actions (such as selecting a vote auditing committee) can usually be achieved using oracles or commitment schemes.

  Our smart contract example will use a random hash in the `generateRandomCommittee` function for committee selection:

```
function generateRandomCommittee() public {
    bytes32 randomHash = keccak256(abi.encodePacked(block.
timestamp, msg.sender));
}
```

Each of these techniques comes with its trade-offs in terms of complexity, cost, and performance. The best combination will depend on the specific requirements of the voting system.

Setting up the blockchain network on Azure

Now, let's progress with setting up the cloud infrastructure for the voting application, which includes a blockchain consortium on Azure and the relevant voting smart contract. The scripts in this section will leverage the Azure **Command-Line Interface** (**CLI**) for execution. We will deploy the smart contract with Truffle, and we will also need MetaMask or a similar Ethereum wallet to sign the voting transactions.

Creating a blockchain consortium on Azure

We will create a blockchain consortium through the Azure portal using bash instructions in the Azure CLI to set up an Ethereum network. Here is an example using the Azure CLI:

```
rg_name="MyResourceGroup"
location="eastus"
consortium_name="MyBlockchainConsortium"
```

First of all, we must set some global variables to indicate the following:

- The resource group where we will save all the deployed resources (the `rg_name` variable).

- The Azure region of deployment (the `location` variable).

- The name of the blockchain consortium being created (the `consortium_name` variable).

The CLI instruction to create a resource group is as follows. Here, we must use the `az group create` command with the resource group name and location as parameters:

```
az group create --name $rg_name --location $location
```

Creating a blockchain consortium requires us to specify the resource group, consortium name, location, and the number of members (at least two) to deploy as part of the consortium. We will use the `az blockchain create` command for this, as follows:

```
az blockchain create --resource-group $rg_name --name $consortium_name
--location $location --consortium-member-count 2
```

Create a voting smart contract

Now that we have a blockchain network up and running, we can proceed with the implementation of the voting logic in a smart contract. Let's create a file named `Voting.sol` and add the following Solidity code to it:

```
pragma solidity ^0.8.0;

contract Voting {
    bytes32[] public candidateList;
    mapping (bytes32 => uint8) public votesReceived;

    constructor(bytes32[] memory candidateNames) {
        candidateList = candidateNames;
    }

    // Vote for a candidate
    function voteForCandidate(bytes32 candidate) public {
```

```
        require(validCandidate(candidate));
        votesReceived[candidate] += 1;
    }


    // Check if a candidate is valid
    function validCandidate(bytes32 candidate) view public returns
(bool) {
        for(uint i = 0; i < candidateList.length; i++) {
            if (candidateList[i] == candidate) {
                return true;
            }
        }
        return false;
    }
}
```

The smart contract does the following:

- Stores a list of candidates in the `candidateList` variable and their vote counts in the `votesReceived` variable.

- Initializes a new instance of the voting contract by passing the candidate's name to the constructor and adding it to the list

Compiling and migrating the smart contract using Truffle

Remember that the complete source code of the smart contract is in this book's GitHub repository. We can now proceed with building and deploying the contract on the provisioned blockchain network. First, we'll start a new Truffle project and place the `Voting.sol` file in the `contracts` directory. To compile the smart contract, we can either use the relevant command from the Truffle dashboard in VS Code, as shown in the following screenshot, or run a simple bash command:

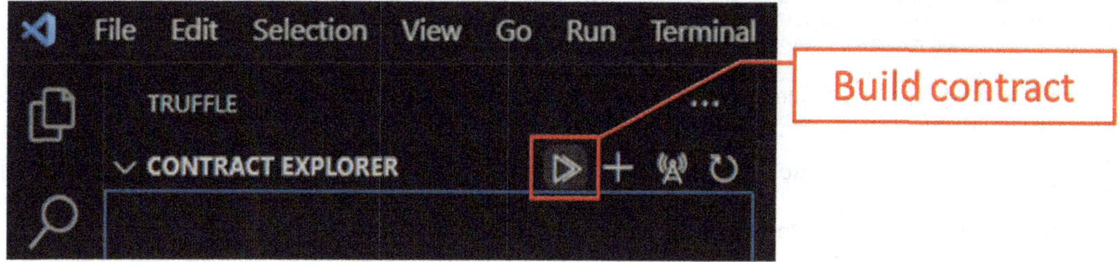

Figure 14.2 – Command to build a smart contract with Truffle in VS Code

The bash command to compile a smart contract in Truffle is as follows:

```
truffle compile
```

Once the contract has been compiled, we need to deploy—or migrate—it to the blockchain network. We'll create a migration script in the `migrations` folder in the Truffle project and run the migration task by executing the following bash command:

```
truffle migrate --network azureNetwork
```

In the preceding command, `azureNetwork` is specified in the `truffle-config.js` file and should have configuration details specific to the blockchain service previously provisioned in Azure.

Developing the voting application frontend

After deploying the contract into the blockchain network, we need a frontend user interface for interacting with the voting application. The frontend code will then communicate with the smart contract and perform voting operations.

We will now create a Blazor UI frontend to interact with the Solidity smart contract. As Blazor runs on .NET, this example will use the Nethereum library, which is a .NET library for integrating Ethereum smart contracts using C# code.

Installing the necessary packages

First, let's make sure we add the `Nethereum.Web3` package to our Blazor project, either from the NuGet package manager in Visual Studio, as illustrated in the following screenshot, or by running a command-line instruction to install the package:

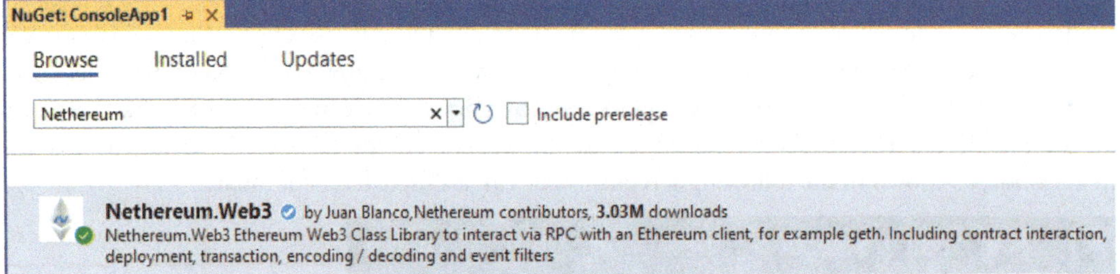

Figure 14.3 – The Nethereum.Web3 NuGet package

The bash command to add the `Nethereum.Web3` package to our .NET solution is as follows:

```
dotnet add package Nethereum.Web3
```

Interacting with the smart contract

Now, we can focus on the interaction with the smart contract by using the Nethereum library. Let's add a new C# class to our project named `EthereumService.cs`. This class will be responsible for all interactions with the Ethereum blockchain.

When instantiated, the `EthereumService` class creates a wallet account for the user with the indicated private key. The `VoteAsync` method is responsible for interacting with the smart contract at the provided blockchain address. Once it's obtained a reference to the `voteForCandidate` function defined in the contract, an asynchronous transaction is initiated between the client code in C# and the on-chain code in Solidity:

> **Secure your private keys**
>
> Never store your private keys in a project configuration file, or even worse, hard code it into your source code. Please consider a more secure secret management solution, such as Azure Key Vault.

```csharp
using Nethereum.Web3;
using Nethereum.Web3.Accounts;
using System.Threading.Tasks;

public class EthereumService
{
    private readonly Web3 web3;
    const string contractAddress = "Contract address";
    const string abi = "contract ABI code";

    public EthereumService(string url, string privateKey)
    {
        var account = new Wallet(privateKey, null);
        web3 = new Web3(account, url);
    }

    public async Task<string> VoteAsync(uint8 candidateIndex)
    {
        var contract = web3.Eth.GetContract(abi, contractAddress);
        var voteFunction = contract.GetFunction("voteForCandidate");
        var transactionInput = await voteFunction.
SendTransactionAsync(candidateIndex);
        return transactionInput;
    }
}
```

Creating a Blazor component for voting

The final touch will involve creating a Blazor component that provides the user interface. We'll call this component `Vote.razor` and implement it in the following Razor code:

```razor
@inject EthereumService EthereumService

@code {
    private int selectedCandidateIndex;
    private string transactionHash;
    private bool hasVoted;

    private async Task VoteAsync()
    {
        transactionHash = await EthereumService.VoteAsync((uint8)
selectedCandidateIndex);
        hasVoted = true;
    }
}
```

Within the `Vote.razor` file, we will include some simple HTML. Feel free to make it as rich and sophisticated as needed:

```razor
@page "/vote"

@if (hasVoted)
{
    <div>
        <h3>You have successfully voted!</h3>
        <p>Your transaction hash is: @transactionHash</p>
    </div>
}
else
{
    <div>
        <h2>Select candidate to vote</h2>
        <select @bind="selectedCandidateIndex">
            <option value="0">Candidate 1</option>
            <option value="1">Candidate 2</option>
            <option value="2">Candidate 3</option>
        </select>
        <button @onclick="VoteAsync">Vote</button>
    </div>
}
```

Running your application

At this point, we should be able to run our Blazor application in a web browser. Navigate to the `/vote` endpoint to interact with the voting app.

> **Please note**
>
> This is a very basic example and doesn't handle several production-ready concerns such as transaction confirmations, error handling, or user authentication. But it should give you a starting point from which you can build a more robust solution.

Before we close this chapter, we'd like to add a brief consideration of future trends in decentralized voting systems, emphasizing emerging technologies and their potential impact on the field.

Future trends in decentralized voting systems

Decentralized voting systems represent a transformative approach to conducting elections, leveraging blockchain technology to ensure transparency, security, and integrity of votes. These systems aim to mitigate traditional challenges such as fraud, coercion, and inefficiency, offering a pathway to more democratic and accessible voting processes.

Blockchain technology and smart contracts are at the core of decentralized voting systems. They ensure that each vote is encrypted and recorded on a distributed ledger, making the voting process tamper-proof and transparent. Smart contracts can automate vote counting and result tabulation, reducing human error and manipulation. As we have learned in this chapter, ZKPs allow voters to prove that their vote is valid within the system's rules without revealing who they voted for. This technology enhances privacy and security, ensuring voter anonymity while maintaining the integrity of the voting process. Similarly, DID systems enable voters to verify their identity securely and privately, without the need for a central authority. This can significantly increase accessibility, allowing for remote voting without compromising security.

In synthesis, decentralized voting systems do the following:

- **Increase accessibility and participation** by making it easier for people to vote, regardless of their location, thereby potentially increasing voter participation and engagement.

- **Enhance security and trust** by leveraging blockchain and associated technologies. By doing this, these systems can significantly reduce the risk of vote tampering and fraud, increasing trust in the electoral process.

- **Address regulatory and legal hurdles** by establishing legal frameworks that accommodate such technology while ensuring the integrity of the electoral process, which is crucial.

Decentralized voting systems, powered by emerging technologies such as blockchain, ZKPs, DID, and potentially quantum computing, hold the promise of revolutionizing the way elections are conducted. They offer a pathway to more secure, transparent, and accessible voting processes. However, addressing scalability, regulatory, and technological challenges is essential for their widespread adoption and success. As these technologies continue to evolve, they will undoubtedly shape the future of democratic processes, making voting more inclusive and resilient against threats.

Summary

This hands-on lab offered a comprehensive guide to building a decentralized voting application while utilizing Azure services for running a blockchain network and hosting a .NET Blazor frontend. The primary goal was to educate participants on how to create a robust, scalable, and secure voting platform that can be transparently audited and managed.

We learned about two key capabilities from this lab. The first one is related to the use of a smart contract for the voting logic, whereas the second one is related to the Blazor frontend and communication with the smart contract.

More specifically, we learned how to write, deploy, and interact with an Ethereum smart contract that handles the voting logic while using the Solidity programming language and the Nethereum library for .NET. We also developed a user interface in Blazor that interacts with the Ethereum smart contract, enabling users to cast votes, view candidates, and see real-time results.

In the next chapter, we will dive into another hands-on lab for creating verifiable digital proofs of ownership of digital assets.

Further reading

- *Get started with Azure*: `https://azure.microsoft.com/en-us/get-started`
- *Blazor*: `https://dotnet.microsoft.com/en-us/apps/aspnet/web-apps/blazor`
- *Nethereum*: `https://nethereum.com/`

15

Creating Verifiable Digital Ownership on GCP

Digital ownership has become a cornerstone of our modern economy. As digital assets such as **Non-Fungible Tokens (NFTs)**, cryptocurrencies, digital real estate, and intellectual property become more prevalent, verifying their ownership in a secure, transparent, and immutable manner has never been more critical.

The advent of digital ownership marks a paradigm shift in how we perceive and interact with assets in the digital realm, profoundly impacting various aspects of our lives. It empowers creators and consumers alike by providing a platform for authenticating, buying, selling, and trading digital goods with a level of security and transparency previously unattainable. This has opened up new avenues for artists, musicians, and writers to monetize their work directly and for collectors and investors to diversify their portfolios into digital art, virtual real estate, and other digital collectibles. Moreover, the use of blockchain technology in establishing proof of ownership and provenance of digital assets such as NFTs not only ensures the integrity and uniqueness of these assets but also offers a new form of engagement and community building around shared interests. As digital ownership continues to weave itself into the fabric of our digital economy, it challenges traditional models of ownership and copyright, prompting legal, economic, and social adjustments to accommodate the burgeoning digital market space.

Welcome to this hands-on lab for creating verifiable credentials for proving ownership of digital assets. In this lab, we'll explore how to establish secure, traceable, and verifiable ownership records for digital assets using blockchain technology. Leveraging the power of **Google Cloud Platform (GCP)**, we'll walk together through the essential tools and services that we need to implement a decentralized application for managing digital proofs of ownership.

In this chapter, we'll cover the following key topics:

- Introduction to verifiable digital ownership
- Setting up a blockchain network on GCP
- Creating verifiable digital ownership records
- Integrating the solution with other GCP services
- Deploying the solution into GCP

Technical requirements

To run the scripts presented in this chapter, we need an account in GCP, which you can create for free, inclusive of $300 worth of credits, from here: `https://cloud.google.com/`

Code examples are provided in the Solidity and JavaScript programming languages. We will be using the following:

- Remix (`https://remix.ethereum.org/`) for developing and testing the smart contracts
- MetaMask or another Ethereum-compatible wallet for interacting with the Ethereum blockchain
- Google Cloud's Ethereum blockchain toolkit for deploying and managing the smart contracts

All the source code is available in the GitHub repository: `https://github.com/PacktPub-lishing/Developing-Blockchain-Solutions-in-the-Cloud`

Introduction to verifiable digital ownership

Verifiable digital ownership refers to a system where digital assets or rights are represented, tracked, and verified using blockchain technology. This concept is crucial in various applications, including cryptocurrencies, NFTs, digital rights management, and more. The key aspects of verifiable digital ownership include the following:

- **Digital assets on blockchain**: Digital assets can be anything from digital art, music, videos, online collectibles, and virtual real estate to cryptocurrencies such as Bitcoin or Ether. These assets are often tokenized, which means they are represented as digital tokens on a blockchain. In the case of unique assets such as digital art, these tokens are often NFTs, meaning each token is unique and cannot be interchanged with another.

- **Ownership and provenance**: The blockchain records who currently owns a digital asset. Each token has an owner, and this information is stored in a way that is transparent and tamper-proof. Blockchain also tracks the history of an asset, including its creation and subsequent transfers. This helps in establishing provenance, verifying authenticity, and ensuring the asset's integrity.

- **Decentralization and trust**: Blockchains are decentralized, meaning that they do not rely on a central authority to verify transactions or ownership. Instead, consensus mechanisms among network participants ensure trust and security. Once recorded on a blockchain, data cannot be altered, which builds trust in the system.

- **Smart contracts for automated verification**: Smart contracts are self-executing contracts with the terms of the agreement directly written into code. They can automate the enforcement of ownership rights, such as transferring ownership upon payment.

Blockchain's transparent yet secure ledger ensures that all participants can verify ownership and transaction history. This leads to an expected reduction of fraud thanks to the difficulty in altering immutable blockchain records, which reduces the risk of counterfeited digital transactions.

There are many domains where verifiable ownership of digital assets is already used at scale, which include (but are not limited to) the following:

- **Art and collectibles**: NFTs have popularized the concept of digital ownership in art and collectibles, allowing artists to sell digital works with clear ownership rights

- **Gaming and virtual real estate**: In gaming, blockchain is used to manage ownership of in-game items and virtual land

- **Intellectual property and licensing**: Blockchain can manage and track licenses for digital content, ensuring creators are compensated for their work

In summary, verifiable digital ownership in blockchain solutions provides a secure, transparent, and efficient way to manage and transfer ownership of digital assets. This technology is revolutionizing how we think about ownership and value in the digital realm.

In the next section, we will start the setup of the cloud resources necessary to build a simple yet effective solution for issuing and verifying ownership of digital assets on a blockchain network.

Setting up the blockchain network on GCP

One of the most important aspects of establishing verifiable digital ownership is the use of **smart contracts**. These contracts will encode the terms of the agreement directly written into code. In this lab, we'll learn how to write and deploy smart contracts to manage and facilitate transactions related to digital ownership records. As consistently done in this book, we will work with the Ethereum platform and Solidity, the most widely used smart contract programming language for Ethereum and compatible blockchain networks.

As a hosting cloud of reference, for this lab, we'll be using GCP. Before we start with Ethereum and GCP, let's define a high-level plan of the tasks that we want to complete in this lab, as shown in the following figure:

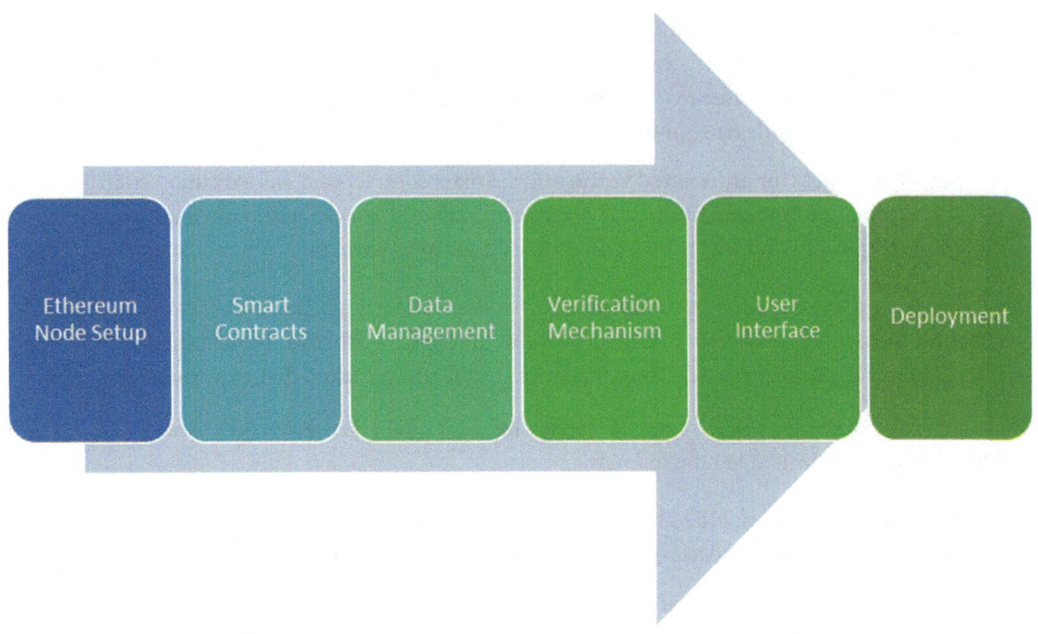

Figure 15.1 – High-level tasks for the verifiable digital ownership solution on GCP

We can detail the sequence of tasks as follows:

1. **Ethereum node setup**: The first step to complete in this lab is setting up an Ethereum node on a GCP VM instance. This forms the backbone of the blockchain infrastructure that we're deploying.

2. **Smart contracts**: We will then write and deploy smart contracts that facilitate asset ownership transactions. The contracts include features for registering, transferring, and verifying ownership.

3. **Data management**: We'll follow the lab's instructions for delving into decentralized data management, exploring both on-chain storage techniques and off-chain solutions such as the **InterPlanetary File System (IPFS)** to manage asset metadata securely and efficiently.

4. **Verification mechanism**: Code examples are provided to demonstrate the creation of verification procedures that are secure and tamper-proof, leveraging blockchain's immutability.

5. **User interface**: Readers will implement a simple yet intuitive frontend using HTML and JavaScript to interact with the deployed smart contract. This ensures a portable and user-friendly way to handle digital asset ownership.

6. **Deployment**: Finally, the lab concludes by providing detailed steps and scripts to deploy the entire verifiable digital ownership solution on GCP, encompassing both backend and frontend components.

Now that we have a plan, let's take action on *Step 1*. Details of how to set up a blockchain node on GCP have already been presented in depth in *Chapter 10* and *Chapter 11*.

Specifically, in *Chapter 10*, we looked at how to configure an Ethereum node using the **Google Kubernetes Engine** (**GKE**), and then interact with it using client tools such as **Geth**.

In *Chapter 11*, we went into the specifics of **Blockchain Node Engine** (**BNE**) on GCP and how to use BNE for running decentralized apps.

For the purpose of this lab, we will use VM instances in GCP for the Ethereum node.

Creating verifiable digital ownership records

Once the hosting infrastructure is sorted, we can proceed with *Step 2*, which is building the smart contract that will do the following:

- Register a new digital asset with unique attributes (for example, ID and ownership details)

- Facilitate the transfer of ownership of a digital asset

- Allow for the verification of the current owner of a digital asset

As mentioned in the technical requirements, we'll be using the Remix IDE for coding the smart contract. Once the contract has been tested, we'll deploy it onto the Ethereum testnet using Google Cloud's Ethereum blockchain toolkit.

The contract exposes functions to add new assets, register a new owner, and transfer ownership. It will also verify the existence of the asset on the blockchain and its current ownership.

The smart contract that we're going to develop is built on the ERC-721, NFT standard (`https://eips.ethereum.org/EIPS/eip-721`), the Ethereum *"request for comment"* that is used for representing ownership of NFTs—that is, where each token is unique. The smart contract performs three tasks (proof of **ownership**, proof of **verification**, and **transfer** of ownership of a digital asset between two parties), as depicted in the following figure:

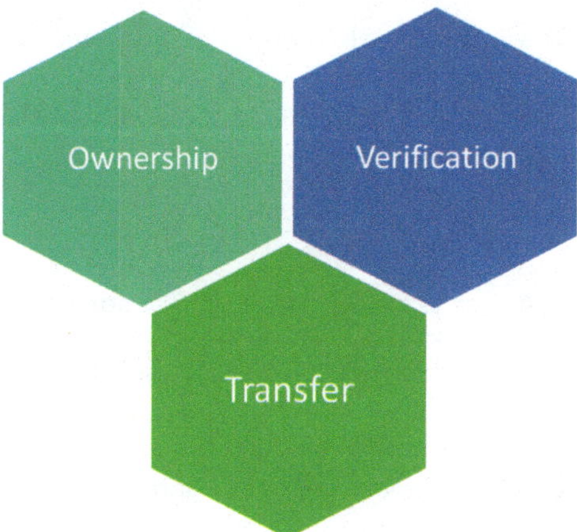

Figure 15.2 – Functionalities of the digital ownership smart contract

A simplified example of a Solidity smart contract that would fit within the context of the lab is given in the following section. The contract allows for the registration and transfer of digital ownership of assets.

The `DigitalAsset` struct defines the structure to hold a digital asset's name and current owner. Each asset is associated with a `DigitalAsset` struct by a unique ID. The `registerAsset` function allows a user to register a new asset by providing a unique asset ID and name. The `transferAsset` function allows the current owner of an asset to transfer ownership to a new owner's address. During the transfer operation, the `AssetTransferred` event is emitted when an asset is transferred from one owner to another. The `verifyOwner` function allows anyone to verify the current owner of an asset by its ID.

To implement this code, open Remix and create a new Solidity file. Copy-paste the full source code from the GitHub repository into the file and compile the code. Later in this chapter, we will see how to deploy the contract to the Ethereum node that we have created in the previous section, and how to interact with it to register assets, transfer them, and verify their ownership.

The following source code represents the `DigitalOwnership` smart contract written in Solidity programming language:

```
contract DigitalOwnership {
    struct DigitalAsset {
        string name;
        address owner;
    }
    mapping(string => DigitalAsset) private assets;


    event AssetTransferred(string assetId, address indexed from,
address indexed to);


    function registerAsset(string memory _assetId, string memory _
name) public {
        require(assets[_assetId].owner == address(0), "Asset already
exists");
        DigitalAsset memory newAsset = DigitalAsset(_name, msg.
sender);
        assets[_assetId] = newAsset;
    }


    function transferAsset(string memory _assetId, address _newOwner)
public {
        require(assets[_assetId].owner == msg.sender, "You are not the
owner");
        assets[_assetId].owner = _newOwner;
        emit AssetTransferred(_assetId, msg.sender, _newOwner);
    }
```

```
    function verifyOwner(string memory _assetId) public view
returns(address) {
        return assets[_assetId].owner;
    }
}
```

Creating secure and tamper-proof verification mechanisms often involves implementing several layers of checks and validations within the smart contract and potentially integrating off-chain services for additional security measures. The following subsections highlight a few advanced Solidity features and practices you can include to enhance the ownership verification mechanisms of your smart contract.

Ownership verification through digital signatures

A powerful method to verify the identity of an asset owner is using digital signatures. This involves creating a hash of the data to be verified and then signing it with the private key of the owner. Anyone with the corresponding public key can then verify the signature and data. We'll now extend the `DigitalOwnership` contract to include signature-based verification.

The `transferAssetWithSignature` function generates a message hash, recovers the signer's address from the signature, and then transfers ownership to the new owner. It also ensures that the signer is the current owner of the asset, and at the end, it emits the `AssetTransferredWithSignature` event to signal to external listeners that an asset has been transferred.

Internally, the signature is calculated with the `recoverSigner` function, which implements an **Elliptic Curve Digital Signature Algorithm** (ECDSA). As usual, the full source code is on the GitHub repository mentioned at the beginning of the chapter.

The following code snippet extends the smart contract with this new functionality:

```
contract DigitalOwnership {
    event AssetTransferredWithSignature(string assetId, address
indexed from, address indexed to, bytes signature);

    function transferAssetWithSignature(string memory _assetId,
address _newOwner, bytes memory _signature) public {
        bytes32 messageHash = keccak256(abi.encodePacked(_assetId,
_newOwner));
        address signer = recoverSigner(messageHash, _signature);
        require(signer == assets[_assetId].owner, "Signature does not
match with the asset's owner");

        assets[_assetId].owner = _newOwner;

        emit AssetTransferredWithSignature(_assetId, msg.sender, _
newOwner, _signature);
    }
```

Multi-signature verification

For added security, we could implement **multi-signature** (**multisig**) functionality. Multisig refers to a verification strategy that requires multiple authorized addresses to confirm a transaction. In this lab, let's work with the assumption that a transfer needs at least two confirmations before being processed. This is defined in the REQUIRED_CONFIRMATIONS constant in the following code:

```
uint8 constant REQUIRED_CONFIRMATIONS = 2;
mapping(string => mapping(address => uint8)) private confirmations;


function confirmTransfer(string memory _assetId, address _newOwner)
public {
    require(assets[_assetId].owner == msg.sender || /* additional
authorized addresses */, "You are not authorized");

    confirmations[_assetId][_newOwner] += 1;

    if (confirmations[_assetId][_newOwner] >= REQUIRED_CONFIRMATIONS)
{
        assets[_assetId].owner = _newOwner;
        emit AssetTransferred(_assetId, assets[_assetId].owner, _
newOwner);
    }
}
```

The new confirmTransfer function will execute the following flow of actions:

1. First, it verifies whether the request for transfer has been submitted by the asset owner or any additional authorized address.

2. Then, it increases the confirmation counter for that asset.

3. Last, when reaching the total required confirmations, it will process the transfer of ownership.

Smart contract deployment

Deploying a smart contract on GCP using the Google Cloud Ethereum blockchain toolkit involves several steps, including the setup of a GKE cluster, as mentioned before.

After deploying an Ethereum node on GKE, we can install the Google Cloud Ethereum blockchain toolkit. This toolkit is maintained by Techlatest.net and a video tutorial on its usage is available at https://techlatest.net/support/ethereum_developer_kit/.

The following helm command will install the Ethereum toolkit on your cluster:

```
helm install my-ethereum ethereum-toolkit/ethereum
```

Now we can prepare the `DigitalOwnership` contract for deployment. First, we need to compile the smart contract by using a Solidity compiler such as `solc`, or a development framework such as Hardhat, to generate the bytecode and the necessary **Application Binary Interface (ABI)**:

```
solc --bin --abi DigitalOwnership.sol -o build
```

At this point, we can create a deployment script to deploy the smart contract to the Ethereum network. Here's an example using JavaScript:

```javascript
const { ethers } = require("hardhat");

async function main() {
    const contractFactory = await ethers.
getContractFactory("DigitalOwnership");
    const contract = await contractFactory.deploy();
    await contract.deployed();

    console.log("DigitalOwnership deployed to: ", contract.address);
}
```

This script creates an instance of the contract factory responsible for connecting to the blockchain network. The `ethers.getContractFactory()` function returns a JavaScript object that represents a smart contract factory. We can use this object to deploy our smart contract by invoking the `deploy()` function. We should make sure that the `DigitalOwnership.sol` source code of the smart contract is in the `contracts` folder in Hardhat.

Off-chain checks

Although this goes beyond smart contract programming, we could also implement additional verification steps that involve off-chain systems. For example, before transferring an asset, a verification token could be sent to the owner's email or mobile number. This token would then need to be provided to complete the transfer (**multi-factor authentication**). These methods, however, would require integrating our smart contract with external systems, typically through oracles or off-chain services.

Overall, by implementing a combination of these techniques, we can make our asset transfer and ownership verification processes both secure and tamper-proof.

Frontend

Creating a frontend UI for users to interact with the decentralized application is crucial for accessibility and usability. **web3.js** and **ethers.js** are commonly used JavaScript libraries that allow dapps to connect to Ethereum nodes, thus enabling users to interact with smart contracts.

We are going to build the web page depicted in the following picture (by all means, please do work on a better-looking UI!):

Manage Your Digital Assets

Register Asset

Asset ID: [____________]
Asset Name: [____________]
[Register]

Transfer Ownership

Asset ID: [____________]
New Owner Address: [____________]
[Transfer]

Verify Ownership

Asset ID: [____________]
[Verify]

Figure 15.3 – A simple web page to interact with the digital ownership smart contract

For this example, let's assume we're using web3.js, HTML, and **vanilla JavaScript**. The UI will allow users to do the following:

- Register a new digital asset

- Transfer ownership of a digital asset

- Verify the current owner of a digital asset

To start, let's create a simple HTML file with input fields for asset registration, ownership transfer, and verification:

```html
<body>
    <h1>Manage Your Digital Assets</h1>

    <h2>Register Asset</h2>
    <form id="registerForm">
        Asset ID: <input type="text" id="registerAssetId" /><br />
        Asset Name: <input type="text" id="registerAssetName" /><br />
        <button type="submit">Register</button>
```

```
        </form>

        <h2>Transfer Ownership</h2>
        <form id="transferForm">
            Asset ID: <input type="text" id="transferAssetId" /><br />
            New Owner Address: <input type="text" id="transferNewOwner"
/><br />
            <button type="submit">Transfer</button>
        </form>

        <h2>Verify Ownership</h2>
        <form id="verifyForm">
            Asset ID: <input type="text" id="verifyAssetId" /><br />
            <button type="submit">Verify</button>
        </form>

        <div id="output"></div>

        <script src="index.js"></script>
    </body>
```

At the bottom of the body of the HTML page, there's a reference to an `index.js` JavaScript file. This file contains the script to handle form submissions and interactions with the smart contract.

First, the script will initialize a `Web3` object. In the real code, please replace the URL with the Ethereum node address in the deployed network:

```
let web3 = new Web3(new Web3.providers.HttpProvider('https://ethereum-
node-address'));
```

Then, let's specify the address and ABI code of the deployed smart contract:

```
const contractAddress = '0x...';
const abi = [...];
```

> **ABI**
>
> An ABI is a JSON representation that tells the interface how to call functions in a smart contract and how data is structured in the contract. After compiling a smart contract with the `solc` compiler, using the `--abi` parameter will generate the ABI code, as in the following example:
>
> **`solc --abi --bin DigitalOwnership.sol -o output`**
>
> Here, the `-o output` parameter specifies the output directory for the generated files.

With that information, we can now create a new Web3 contract instance:

```
const contract = new web3.eth.Contract(abi, contractAddress);
```

After the initialization is done, we can handle the various form interactions. The following diagram shows the interaction flows:

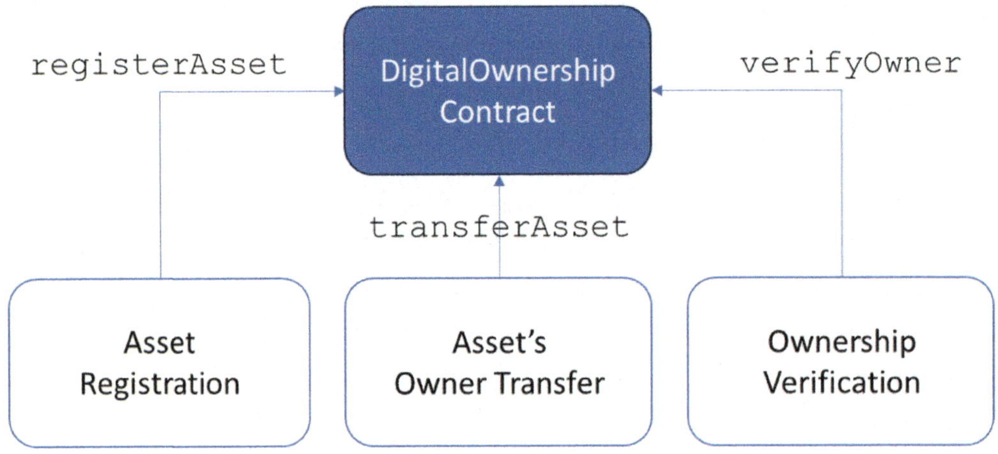

Figure 15.4 – Smart contract interaction flow with the frontend

First is the asset registration, which invokes the `registerAsset` method on the deployed contract:

```
document.getElementById('registerForm').addEventListener('submit',
function(e) {
    e.preventDefault();
    const assetId = document.getElementById('registerAssetId').value;
    const assetName = document.getElementById('registerAssetName').
value;
    contract.methods.registerAsset(assetId, assetName)
    .send({ from: web3.currentProvider.selectedAddress })
    .then((result) => {
        document.getElementById('output').innerHTML = 'Asset
Registered Successfully!';
    })
    .catch((error) => {
        document.getElementById('output').innerHTML = 'Asset
Registration Failed!';
    });
});
```

We then continue with the asset transfer handler by invoking the `transferAsset` function in the smart contract:

```
document.getElementById('transferForm').addEventListener('submit',
function(e) {
    e.preventDefault();
    const assetId = document.getElementById('transferAssetId').value;
    const newOwner = document.getElementById('transferNewOwner').
value;
    contract.methods.transferAsset(assetId, newOwner)
    .send({ from: web3.currentProvider.selectedAddress })
    .then((result) => {
        document.getElementById('output').innerHTML = 'Asset
Transferred Successfully!';
    })
    .catch((error) => {
        document.getElementById('output').innerHTML = 'Asset Transfer
Failed!';
    });
});
```

And finally, the handler for ownership verification invokes the `verifyOwner` function on the contract:

```
document.getElementById('verifyForm').addEventListener('submit',
function(e) {
    e.preventDefault();
    const assetId = document.getElementById('verifyAssetId').value;
    contract.methods.verifyOwner(assetId).call()
    .then((result) => {
        document.getElementById('output').innerHTML = `The owner of
the asset is ${result}`;
    })
    .catch((error) => {
        console.error(error);
        document.getElementById('output').innerHTML = 'Failed to
Verify Owner!';
    });
});
```

In this section, we have created a smart contract that handles all the logic for issuing, transferring, and verifying ownership of digital records. But smart contracts don't run in isolation. We need a mechanism for interacting with them and for storing data, either on the blockchain itself (*on-chain*), or in another centralized digital storage facility (*off-chain*). This can be achieved by integrating additional cloud services with the solution prepared so far, as we'll see in the next section.

Integrating verifiable digital ownership with other GCP services

Managing digital assets in a decentralized manner generally involves storing data on the blockchain and providing functionalities to interact with that data. Blockchain's immutable nature ensures the integrity and reliability of the data. In the context of Ethereum, this data management usually takes place within smart contracts, which may be complemented by decentralized storage solutions such as IPFS for handling large data or files. Let's discover a few ways of storing and managing digital assets in a decentralized manner.

On-chain storage using smart contracts

In Ethereum, the most straightforward way to manage digital assets is by storing all necessary information within a smart contract. The following example extends the `DigitalOwnership` smart contract to include additional attributes, such as metadata, for each digital asset. Metadata can be in any form, for example, a JSON string or IPFS hash. The `registerAsset` function is extended to accept a `string memory _metadata` input parameter. A new `getAsset` function is introduced in the smart contract to retrieve the stored digital asset details:

```
contract DigitalOwnership {
    struct DigitalAsset {
        string name; // Name of the asset
        address owner; // Current owner of the asset
        string metadata; // Additional information
    }


    mapping(string => DigitalAsset) private assets;

    function registerAsset(string memory _assetId, string memory _
name, string memory _metadata) public {
        require(assets[_assetId].owner == address(0), "Asset already
exists");

        DigitalAsset memory newAsset = DigitalAsset(_name, msg.sender,
_metadata);
        assets[_assetId] = newAsset;
    }


    function getAsset(string memory _assetId) public view
returns(DigitalAsset memory) {
        return assets[_assetId];
    }
}
```

Off-chain storage with on-chain references

For managing larger datasets or files, a smart contract is no longer suitable for storage, due to limitations in the blockchain itself and higher gas costs. At that point, we may opt for a decentralized file storage system such as IPFS. With this approach, the smart contract stores a hash reference to the data stored on IPFS.

Let's introduce yet another function, called `registerAssetWithIPFS`, to our smart contract, needed for registering a new digital asset with IPFS metadata:

```
function registerAssetWithIPFS(string memory _assetId, string memory
_name, string memory _ipfsHash) public {
    require(assets[_assetId].owner == address(0), "Asset already
exists");

    DigitalAsset memory newAsset = DigitalAsset(_name, msg.sender,
_ipfsHash);
    assets[_assetId] = newAsset;
}
```

We may also want to handle collections of assets for each user. The `addUserAsset` function represents an example of using an array and mapping an asset to a specific user. Its counterpart, `getUserAssets`, returns the entire collection of assets owned by the given user:

```
mapping(address => string[]) private userAssets;

function addUserAsset(string memory _assetId) internal {
    userAssets[msg.sender].push(_assetId);
}

function getUserAssets(address _owner) public view returns(string[]
memory) {
    return userAssets[_owner];
}
```

Both on-chain and off-chain techniques are perfectly viable, and both can be combined to create a robust system for managing digital assets in a decentralized environment. This will allow a dapp to handle both the metadata and actual content of digital assets while benefiting from the immutability and security features of blockchain technology.

Deploying verifiable digital ownership on GCP

We're at the final stage of our solution. Deploying a verifiable digital ownership solution on GCP involves several steps. This comprehensive guide outlines how to set up the required components, deploy the smart contract, and run the frontend application.

As we have already deployed the cloud infrastructure in a previous step, we now focus on the deployment of the smart contract:

1. **Compile** the Solidity smart contract, if not already done, with the `solc` command:

    ```
    solc --bin --abi -o ./output/path/DigitalOwnership.sol
    ```

2. **Deploy** the compiled contract using web3.js in a Node.js script (let's call it `deploy.js`) to deploy the compiled contract. Make sure that the RPC port of the VM with the Ethereum node is accessible, or use an SSH tunnel to route the request:

    ```javascript
    const Web3 = require('web3');
    const web3 = new Web3(new Web3.providers.HttpProvider("https://
    ethereum-node-address"));
    const abi = [...];  // Contract ABI
    const bytecode = "0x...";  // Compiled contract bytecode

    const contract = new web3.eth.Contract(abi);
    contract.deploy({
        data: bytecode
    })
    .send({
        from: 'your-ethereum-address',
        gas: 1500000
    })
    .then((newContractInstance) => {
        console.log(`Contract deployed at ${newContractInstance.
    options.address}`);
    });
    ```

3. **Run** the deployment script:

    ```
    node deploy.js
    ```

4. **Upload** the frontend files to App Engine (as it's static content, cloud storage would work too):

    ```
    cd path/frontend
    gcloud app deploy
    ```

5. **Update** the frontend JavaScript code to point to the deployed smart contract address and the Ethereum RPC URL, as indicated previously.

> **Scripts for automation**
>
> To automate these steps, we can write shell scripts or use **Infrastructure as Code (IaC)** tools such as Terraform to provision and configure resources on GCP. We could also use GCP's Deployment Manager to create templates for automating the entire setup.

That's it! We've now deployed a verifiable digital ownership solution on GCP using blockchain technology. Well done!

Summary

In this hands-on lab, we explored the fundamentals of creating verifiable digital proofs of ownership of digital assets using blockchain technology on GCP. The lab is designed as an end-to-end tutorial, offering guided steps and code samples that cover key aspects such as Ethereum node setup, smart contract development, decentralized data management, and frontend application integration.

By completing this lab, we have gained a comprehensive understanding of how to build and deploy a complete blockchain-based solution for verifiable digital ownership. This knowledge is invaluable for anyone looking to implement secure, decentralized systems for asset management.

This chapter completes the three hands-on labs in the book. In the next and final chapter, we'll draw some conclusions from this journey, and make some assumptions on the future of cloud-native blockchain solutions

Further reading

Google Cloud's Ethereum blockchain toolkit: `https://console.cloud.google.com/marketplace/product/techlatest-public/ethereum-developer-kit`

16

The Future of Cloud-Native Blockchain

As we stand at the confluence of cloud computing and blockchain technology, the emergence of cloud-native blockchain solutions offers a new frontier in scalable, secure, and easily deployable digital infrastructure. Businesses and organizations are increasingly looking for agile and cost-effective methods to deploy blockchain, and major cloud providers such as AWS, Azure, and GCP are stepping up to fulfill this need. This chapter delves into the evolving landscape of cloud-native blockchain implementations, focusing on the offerings and future directions within AWS, Azure, and GCP platforms.

The convergence of cloud and blockchain presents an unparalleled opportunity for enterprises. On one hand, cloud platforms offer the advantages of high availability, scalability, and global reach. On the other hand, blockchain technology promises unprecedented security and transparency, features that are increasingly necessary in a digitized, interconnected world. But how do these technologies evolve together within the ecosystems of the world's leading cloud providers?

In the following sections, we will explore the current state-of-the-art solutions, scrutinize the unique advantages and challenges posed by each cloud platform, and project future developments that could further redefine the landscape of blockchain technology. Whether you are an enterprise decision-maker, a blockchain developer, or a technology enthusiast, this chapter aims to provide you with insightful perspectives on the rapidly evolving domain of cloud-native blockchain.

Why cloud-native blockchain?

Before diving into specifics, let me elucidate what **cloud-native** actually means in the context of blockchain, why it's an evolutionary step, and how it's different from traditional blockchain deployments. In a nutshell, cloud-native blockchain is about leveraging cloud-specific features such as microservices, containers, serverless computing, and orchestration tools.

Cloud-native blockchain solutions are specifically designed to take advantage of the unique features and benefits offered by cloud platforms. This allows them to achieve greater agility, scalability, and cost-efficiency compared to traditional blockchain deployments. The following figure shows cloud-native architectures as the intersection of four core pillars: containers, microservices, serverless, and automation:

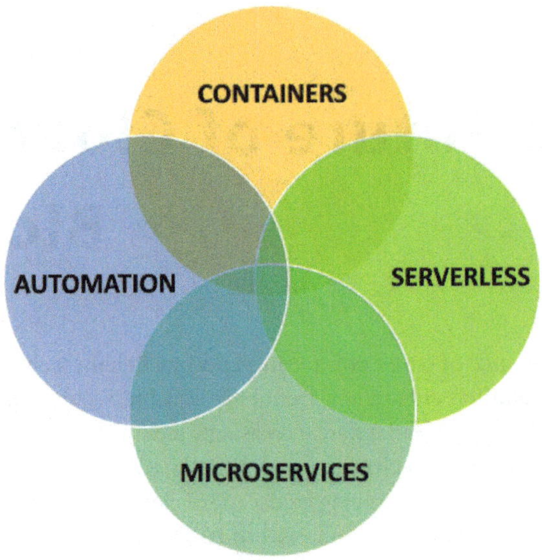

Figure 16.1 – The core pillars of cloud-native architectures

In this book, we looked at cloud **container platforms** for the deployment of the microservice infrastructure required to run blockchain solutions. Containers offer isolation and the portability of services, ensuring secure and predictable execution. This allows for easy deployment across different cloud environments and hybrid deployments. Containers are by their nature lightweight and start up quickly, facilitating agile development and rapid scaling in response to changing demands.

Containers are also ideally placed to support **microservices architectures**, and this modular approach also fits well with the decentralized and distributed characteristics of blockchain-based solutions. Blockchain features can be broken down into independent, loosely coupled microservices, which enable faster development, easier deployment, and improved scalability. Each microservice can be scaled independently based on demand, optimizing resource utilization. Failure in one microservice doesn't affect the entire system. This enhances fault tolerance and ensures continuous operation.

Another significant cloud-native advantage is **serverless computing** that enables pay-per-use models: *Serverless* means removing the need for dedicated servers, reducing infrastructure costs, and eliminating management overhead. Blockchain functions can be triggered by events, optimizing resource utilization and reducing operational complexity. Scalability is then obtained **on demand**: Serverless status automatically scales resources up or down based on workload, ensuring the efficient use of resources and cost savings.

The last pillar of cloud-native architectures is **automated deployment and management**: Tools such as Kubernetes automate the deployment, scaling, and configuration of microservices and containers. This simplifies managing complex blockchain deployments and ensures consistent operation. High availability and fault tolerance are immediate benefits; orchestration tools can automatically restart failed containers or deploy new instances, ensuring the high availability and resilience of the blockchain network.

The cloud offers additional benefits, too:

- **Cloud security**: Cloud platforms offer robust security features and compliance certifications, enhancing the overall security posture of blockchain solutions.

- **Analytics and monitoring**: Cloud tools provide centralized logging and monitoring capabilities, facilitating a better understanding of blockchain network performance and identifying potential issues.

- **Global reach**: Cloud providers offer geographically distributed data centers, enabling blockchain solutions to be deployed and accessed from anywhere in the world.

In this book, we looked at the three major public cloud service providers, AWS, Azure, and GCP, and offered an insight into how these three players support the development and deployment of blockchain solutions and what additional cloud services can be leveraged to enhance the experience of cloud-native decentralized apps.

Blockchain for good

In an era marked by global interconnectedness and technological innovation, the quest for improving the human condition has taken on unprecedented dimensions. With each passing day, new challenges arise, necessitating novel solutions that transcend traditional boundaries. In this pursuit, blockchain technology has emerged as a revolutionary force with immense potential to reshape the landscape of humanitarian efforts.

Originally conceptualized as the underlying technology supporting the cryptocurrency phenomenon, blockchain has swiftly evolved into a versatile tool with applications that transcend the financial realm. Its decentralized and immutable nature has captured the attention of visionaries, technologists, and humanitarians alike, who recognize its power to foster transparency, accountability, and efficiency in humanitarian endeavors.

At its core, humanitarian work is focused on addressing the needs of vulnerable populations, providing aid during crises, and empowering communities to thrive. However, despite the noble intentions behind these initiatives, traditional systems often grapple with challenges such as corruption, bureaucratic inefficiencies, and difficulties in tracking the flow of resources. It is within this context that blockchain technology offers a beacon of hope, promising a transformative shift towards a more equitable and impactful approach to humanitarian aid.

The dynamic landscape of blockchain technology has a great potential application in the realm of humanitarianism. From disaster response and refugee assistance to ensuring fair aid distribution and fostering trust among stakeholders, blockchain's unique attributes hold the promise of overcoming long-standing obstacles in the humanitarian sector.

What are the fundamental principles that underpin blockchain technology? Which highlighted features make it a game-changer for humanitarian purposes? Are there any real-world use cases where blockchain has already made significant strides in improving the effectiveness and accountability of humanitarian initiatives?

As we embark on this exploration of blockchain's utilization for good humanitarian purposes, it is crucial to maintain a balanced perspective, acknowledging both its transformative capabilities and inherent limitations. While blockchain technology is not a panacea for all humanitarian challenges, it undoubtedly represents an innovative approach that, when applied judiciously, can revolutionize the way we deliver assistance and support to those most in need.

Ultimately, this section seeks to ignite a dialogue surrounding the fusion of cutting-edge technology and compassionate altruism. By understanding how blockchain can serve as an enabler of positive change, we aspire to inspire future generations of innovators, humanitarians, and policymakers to collaborate toward a world where the full potential of technology is harnessed for the greater good of humanity.

Identity verification and protection

In the digital identity context, blockchain can offer a secure and reliable way of managing and verifying identities, especially for refugees and displaced persons who may lack access to traditional forms of ID. Projects such as ID2020 (`https://www.id2020.org/`) are exploring blockchain's potential to provide digital identities to the millions worldwide who are without them. Such initiatives can significantly impact access to services, including banking, healthcare, and voting, ensuring inclusivity and support for vulnerable populations.

Charitable donations and aid distribution

Blockchain technology can increase transparency and traceability in charitable donations, ensuring that funds reach their intended destinations without diversion or corruption. Organizations such as **GiveTrack** (`https://www.givetrack.org/`), a platform created by BitGive, use blockchain to allow donors to trace their contributions in real time and see exactly how their money is being used. Similarly, the World Food Program's **Building Blocks** project (`https://www.wfp.org/building-blocks`) uses blockchain to distribute cash assistance to those in need securely and efficiently, reducing transaction costs and potential fraud.

The digital archiving and provenance of cultural artifacts

Blockchain can be employed to create immutable digital records of cultural artifacts, artworks, and historical documents. This application not only helps prove the authenticity and ownership of such items but also preserves the knowledge and context surrounding them, which might otherwise be lost over time. By leveraging blockchain, applications can enable the verification of an item's history and provenance, enhancing transparency and trust in the art market. Moreover, they would allow for the digital preservation of cultural heritage, making it accessible to future generations without the risk of tampering or loss.

An example of this use case is the **Ukrainian Heritage Hub** (`https://www.heritagehub.org/`). Amidst the challenges posed by ongoing conflicts, blockchain technology is being explored as a means to document Ukrainian cultural artifacts, monuments, and historical sites securely and immutably. The Heritage Hub aims to use blockchain for cataloging and preserving digital records of cultural and historical significance. By creating a decentralized ledger of cultural items, these initiatives ensure that even if physical artifacts are lost or destroyed, digital representations, complete with historical data, provenance, and significance, are preserved for future generations.

Blockchain's application in this area can involve creating detailed digital twins of artifacts, buildings, or sites using 3D scanning technology, which are then recorded on the blockchain. This not only aids in preservation efforts but can also assist in restoration and educational projects, allowing people worldwide to access and learn about Ukrainian culture in an interactive and engaging manner.

Such efforts highlight the power of blockchain as a tool for cultural preservation, ensuring that the richness of Ukraine's heritage is maintained and accessible in the digital realm, regardless of the challenges faced in the physical world.

Role of cloud providers in shaping the future of cloud-native blockchain

The role of cloud service providers in shaping the future of cloud-native blockchain cannot be overstated. These platforms are doing more than just hosting blockchain networks; they are actively contributing to how businesses and organizations will interact with and utilize blockchain technologies for years to come. We can think of at least six key benefits that cloud providers can bring to blockchain-based applications, and we're sure there are more. Each benefit is a slice of the big cake that cloud-native represents, as in the next figure. None is larger or more important than the others; all contribute to the success of the solution.

Figure 16.2 – Benefits to blockchain solutions by cloud providers

Let's start with the **standardization and simplification** of blockchain services and protocols.

AWS, Azure, and GCP are in a unique position to standardize blockchain deployments. With a variety of blockchain frameworks and protocols available, having a few influential players standardize the technology can aid mass adoption. Through managed services such as Amazon Managed Blockchain, Azure Confidential Ledger, or Hyperledger Fabric on GCP, these providers simplify the complex task of setting up and maintaining a blockchain network.

The overall **governance of a blockchain network** is also simplified. Managing a consortium blockchain, which is commonly used in enterprise settings, involves complex governance. Azure, for example, provides built-in governance features to manage access and validation nodes in the network, as seen in *Part 3* of this book.

All cloud providers make it easy to **integrate their blockchain services** with other cloud services for off-chain data storage, event messaging, data analytics, machine learning, and even IoT. The integration of these services with blockchain applications is vital for creating more robust, scalable, and versatile solutions, as we have seen with the Amazon Quantum Ledger Database in *Chapter 6* and Google BigQuery in *Chapter 12*.

Accessibility also means the democratization of user-friendly interfaces. Amazon, Microsoft, and Google, in their cloud offerings, are democratizing access to blockchain by offering easy-to-use interfaces for setting up and managing blockchain networks. This reduces the entry barrier for businesses interested in adopting blockchain but lacking the necessary expertise. These platforms also provide extensive training resources, documentation, and community support. This helps in educating the workforce and thus broadening the base of people capable of implementing blockchain solutions.

We have already mentioned **global reach** as an extra technical benefit of deploying a dapp in the cloud. Global reach is more than just deployment in multiple geographic regions, though. Yes, this is certainly a key benefit, as it immediately ensures that cloud-native blockchain solutions are viable on a global scale and can comply with local regulations. Global reach is also about future-proofing the technology by expanding your horizons, marketing, and talent outside of your local data center. For example, all three cloud providers invest heavily in R&D, pushing the boundaries of what is possible with blockchain technology. For instance, they are exploring issues such as scalability, interoperability, and new consensus algorithms, shaping how cloud-native blockchain evolves.

They are continuously **fostering innovation** and the creation of new ecosystems through partner networks: AWS, Azure, and GCP offer marketplaces where third-party vendors can offer blockchain solutions, thereby nurturing an ecosystem of developers and business solutions. Especially in the case of Azure and GCP, contributions to open source blockchain projects are encouraged and help evolve the technology and build community trust.

In summary, AWS, Azure, and GCP are not just service providers; they are trendsetters in the blockchain space. Their platforms act as both a testing ground for new ideas and a launchpad for mature solutions, shaping the future of how businesses and organizations will leverage cloud-native blockchain technology.

Challenges and opportunities for cloud-native blockchain

The journey towards the mass adoption of cloud-native blockchain technologies presents a mixed bag of challenges and opportunities. The convergence of cloud computing and blockchain technologies provides fertile ground for innovation, but there are key hurdles to overcome. The following figure illustrates the most common challenges in the adoption of cloud-native blockchain at scale:

Figure 16.3 – Challenges in the adoption of cloud-native blockchain at scale

One of the significant challenges lies in the **interoperability** of different chains, that is, making various blockchain protocols work seamlessly with each other and with existing systems. While cloud providers such as AWS, Azure, and GCP are developing solutions, full interoperability remains an elusive goal.

Compliance is complex, as blockchain applications often deal with sensitive data, and navigating the global landscape of data protection regulations is challenging. Cloud providers have to ensure their blockchain services comply with laws such as GDPR, HIPAA, and others.

Probably the most common challenge in layer 1 blockchain networks is **scalability**. Traditional blockchains struggle with scalability issues, and while cloud-native approaches promise improvements and high-throughput, low-latency solutions are still under development.

This leads us to consider **costs** carefully. While cloud-native solutions offer scalability, they are not necessarily cheaper, especially for smaller organizations. Understanding the total cost of ownership, which includes not just cloud fees but also the costs of migration, management, and any necessary modifications to the existing systems, is crucial.

Blockchain is generally secure, but integrating it into existing cloud infrastructures could create potential vulnerabilities. Both the blockchain and cloud components have to be secure to ensure the overall system's **security**.

Lastly, we shouldn't underestimate that blockchain technology is still in its emerging phase, with quick changes happening regarding the complexity of setting up, maintaining, and effectively using it productively. This requires specialized skills. Although cloud providers offer simplified solutions, a **skill gap** remains.

Obviously, the opportunities for cloud-native blockchain solutions are equally important, and as we hope the readers have appreciated in this book, the ease of deployment and management that clouds offer are a competitive advantage over non-cloud-based approaches. The following figure illustrates the most common opportunities for cloud-based solutions:

Figure 16.4 – Opportunities for cloud-native blockchain solutions

Cloud-native blockchain solutions are far easier to deploy and manage than traditional, on-premises blockchain networks. **Managed services** offered by AWS, Azure, and GCP provide one-click solutions to set up blockchain infrastructures, which could speed up adoption rates.

The cloud brings **innovation** at scale: cloud-native architecture enables scalable blockchain solutions. Cloud providers' global infrastructure allows blockchain networks to grow and contract dynamically based on demand.

We also mentioned the opportunity to **integrate** blockchain applications with existing cloud services such as AI, IoT, and data analytics that can result in powerful, versatile applications. This is especially true in fields such as supply chain management, healthcare, and finance.

If data governance is a challenge, providing **transparency** on how data is used is an opportunity. By implementing blockchain within cloud services, businesses can benefit from enhanced data governance capabilities. Blockchain's immutable ledger technology, combined with the cloud's scalability, offers unparalleled data tracking and auditing capabilities.

Global reach means regional compliance, too. Cloud providers have data centers worldwide, allowing for the easy global deployment of blockchain networks that are also in compliance with local regulations. This opens doors for multinational businesses and supply chains.

Global reach also has an impact on the development of global communities and partner **ecosystems**. The growth of cloud-native blockchain technologies will likely be accompanied by an expanding ecosystem of tools, libraries, and community contributions that make the technology more robust and accessible.

Let's take an example of how balancing these challenges and opportunities has the potential to revolutionize various sectors: A **connected and automated supply chain**. The potential benefits of implementing blockchain and IoT technologies in the manufacturing sector are no secret, especially in enhancing the security and transparency of supply chains. For those unfamiliar with this area, it's important to note that traditional supply chain management often faces challenges in reliability, measurement accuracy, and process transparency among all involved parties.

Blockchain and IoT technologies offer promising solutions to these issues. IoT devices can automate metric collection at any point in the supply chain, while a decentralized digital ledger provides a secure and unchangeable record of transactions. Furthermore, blockchain-based smart contracts can apply rule-based logic to validate these data, ensuring each stage of the supply chain is updated and visible to all parties in a trustworthy and clear manner. For instance, a smart contract can embed specific terms, conditions, and logic to verify the proper transfer of goods between different supply chain phases.

The author worked on a blockchain solution for the supply chain of dairy products. A summary of the experience is described in the *Secure Your Supply Chain with the Azure IoT and Blockchain Cloud* article, available at `https://learn.microsoft.com/en-us/archive/msdn-magazine/2019/august/blockchain-secure-your-supply-chain-with-the-azure-iot-and-blockchain-cloud`.

Have a look at the following figure, which simplifies the distribution process in the supply chain. Here, a food producer sends products to a food processing company for packaging and distribution. The food items are packaged under specific conditions of temperature and humidity, and it's crucial that similar conditions are also maintained during their transportation to both the warehouse and the retail outlet:

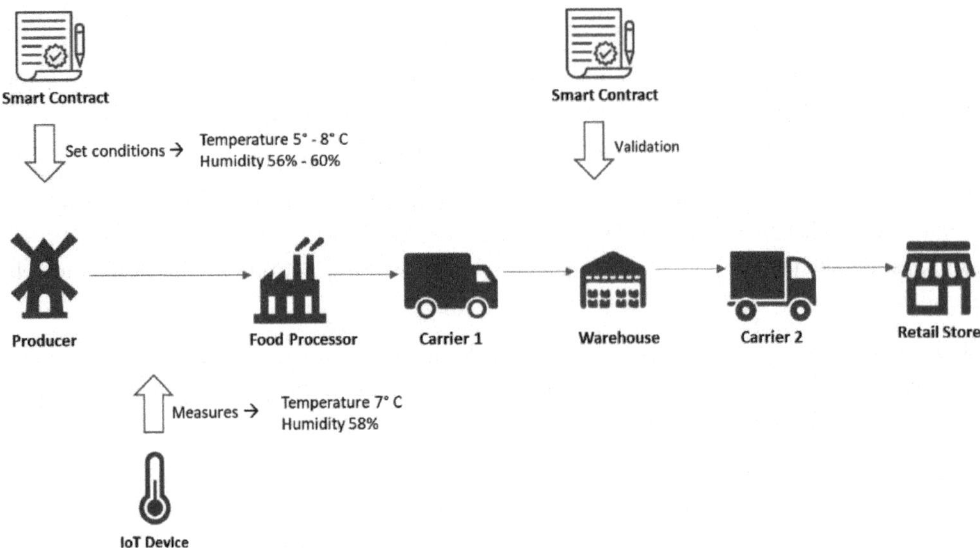

Figure 16.5 – IoT and blockchain-connected supply chain

In this setup, specific conditions are programmed into a blockchain smart contract, and IoT devices equipped with temperature and humidity sensors are placed in containers and vehicles used for storing and transporting the goods. These devices continuously record temperature and humidity data at set intervals throughout the storage and distribution stages. The collected data are then compared with the predefined standards in the smart contract. If the recorded temperature or humidity levels breach these preset thresholds, the transportation status of the goods is updated to **non-compliant** in the blockchain's digital ledger. This change in status is logged and becomes visible to all supply chain participants.

Moreover, each party responsible for storing or transporting the goods is required to designate the next handler in the supply chain, updating this information in the system. The IoT devices continue to send telemetry data to a central IoT hub. This process enables both the initiator and all participants of the supply chain to pinpoint exactly which party failed to adhere to the compliance standards if the temperature or humidity levels were not maintained at any point in the chain.

The moral of the story is that the involvement of major cloud service providers is likely to accelerate the adoption of blockchain technology at scale, offering businesses more robust, scalable, and easy-to-deploy blockchain solutions.

Final thoughts and predictions for the future of cloud-native blockchain

As we look toward the horizon of cloud-native blockchain technology, the prospects are both promising and complex. While the union of cloud computing and blockchain technology has laid the groundwork for next-generation digital infrastructure, the path to realizing its full potential is fraught with both technical and regulatory hurdles. However, with major cloud service providers such as AWS, Azure, and GCP at the helm, the rate of innovation and solution deployment is accelerating.

The authors, reviewers, and the publisher of this book would like to thank the readers for sharing this journey with us. We leave you with our final thoughts and predictions. Let us know yours!

Final thoughts

Our final thoughts are illustrated in the following figure:

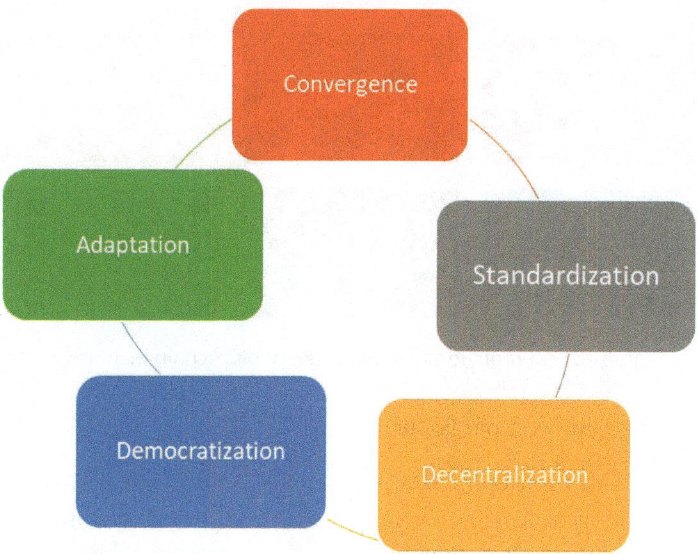

Figure 16.6 – Final thoughts on blockchain opportunities

We have seen many bulleted lists in this book. We promise this is the last one:

- **Convergence of technologies**: The amalgamation of blockchain with other emerging technologies such as AI, IoT, and edge computing, especially within cloud ecosystems, is likely to create holistic solutions that are greater than the sum of their parts.

- **Standardization**: As the technology matures, expect more standardization, especially with cloud giants driving the market. Standardized APIs, protocols, and governance models will make adoption easier and more widespread.

- **Decentralization vs. centralization**: A key question for the future is how the decentralized ethos of blockchain will mesh with somewhat centralized cloud services. This duality could define new hybrid architectures that capture the best of both worlds.

- **Democratization of blockchain**: Cloud-native architectures and managed services will likely democratize access to blockchain technology, making it accessible to businesses and developers of all sizes and scopes.

- **Industry-specific adaptations**: As the technology matures, it is likely to branch into industry-specific solutions tailored to meet the regulatory and operational needs of sectors such as healthcare, finance, and logistics.

Predictions

Now, let's open our crystal ball. It'd be interesting to read this section in 1, 3, and 5 years from now and see how many of these predictions were actually true! The following last colorful figure shows them all in a single image:

Figure 16.7 – Predictions for the future of blockchain technology

Did we say our final thoughts were our last list?

- **Interoperability**: Efforts towards achieving full interoperability between various blockchain protocols and existing systems will likely intensify, perhaps even resulting in industry-standard protocols that enable seamless communication between different blockchain networks.

- **Store of value**: The potential evolution of Bitcoin from a store of value to a platform supporting **Decentralized Finance (DeFi)** applications through the integration of Layer 2 solutions. Layer 2 protocols, such as Stacks, could enable Bitcoin to support smart contracts and DeFi functionalities, thus broadening its use cases beyond just being a digital currency.

- **Scalable consensus algorithms**: The need for more scalable blockchain networks will drive the development of new consensus algorithms that can handle higher transaction volumes while maintaining security and decentralization.

- **Regulatory clarity**: As blockchain becomes more mainstream, we can expect clearer regulatory guidelines that will help businesses navigate compliance more easily.

- **Data privacy and sovereignty**: Given global concerns about data privacy, we might see new blockchain-based data sovereignty solutions that allow individuals and entities to own and control their data while still benefiting from cloud-based analytics and machine learning tools.

- **Sustainable blockchain**: With growing awareness of the environmental impact of blockchain technologies such as Bitcoin, future cloud-native blockchain solutions will likely focus on sustainability, possibly utilizing greener consensus algorithms or carbon offset programs.

In conclusion, the union of blockchain and cloud computing heralds a paradigm shift in how we think about digital transactions, data governance, and computational trust. While the journey ahead is complex, the rewards—ranging from enhanced transparency and security to global and scalable digital ecosystems—are incredibly promising. The role of major cloud service providers in shaping this future is undeniably central, as they act as both catalysts and gatekeepers in this evolving landscape.

What we're observing is that the current rise of blockchain projects resembles the dot-com boom in the late 1990s, with a surge in new companies of varying viability. Like the dot-com era, only a few blockchain endeavors may persist and significantly impact the industry, highlighting the importance of strong fundamentals, innovation, and practicality for long-term success.

Summary

This final chapter delved into the intricacies and possibilities of cloud-native blockchain, particularly focusing on the roles and offerings of the major cloud service providers AWS, Azure, and GCP. Starting with an exploration of what cloud-native means within the blockchain context, the chapter discussed how these giants of cloud computing are standardizing, simplifying, and democratizing blockchain technologies. It also touched on the challenges that lie ahead, such as interoperability, scalability, and regulatory compliance, while highlighting the immense opportunities for innovation, integration, and global reach. Finally, the chapter closed with thoughtful predictions for the future of cloud-native blockchain, including its convergence with other technologies, regulatory clarity, and evolution into industry-specific solutions.

Stay in touch! The journey is not over with this book; it's actually just the beginning of it. There are many ways you can deepen your understanding of blockchain technology, starting with additional books, videos, courses, and conferences by Packt. Read more about all these opportunities at https://www.packtpub.com/.

As we stand at the intersection of cloud and blockchain, let's remember that every intersection is not just a point of meeting but a point of transformation. It's not merely where two paths cross; it's where they blend to forge a trail that didn't exist before. And it's up to us to navigate and shape that new trail toward a future filled with promise and possibility.

Immensely grateful,

Stefano and Michael.

Index

packtpub.com

Subscribe to our online digital library for full access to over 7,000 books and videos, as well as industry leading tools to help you plan your personal development and advance your career. For more information, please visit our website.

Why subscribe?

- Spend less time learning and more time coding with practical eBooks and Videos from over 4,000 industry professionals

- Improve your learning with Skill Plans built especially for you

- Get a free eBook or video every month

- Fully searchable for easy access to vital information

- Copy and paste, print, and bookmark content

Did you know that Packt offers eBook versions of every book published, with PDF and ePub files available? You can upgrade to the eBook version at packtpub.com and as a print book customer, you are entitled to a discount on the eBook copy. Get in touch with us at customercare@packtpub.com for more details.

At www.packtpub.com, you can also read a collection of free technical articles, sign up for a range of free newsletters, and receive exclusive discounts and offers on Packt books and eBooks.

Other Books You May Enjoy

If you enjoyed this book, you may be interested in these other books by Packt:

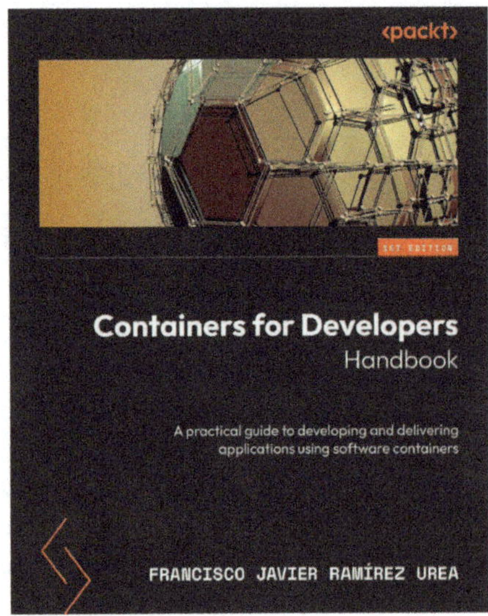

Containers for Developers Handbook

Francisco Javier Ramírez Urea

ISBN: 978-1-80512-798-7

- Find out how to build microservices-based applications using containers
- Deploy your processes within containers using Docker features
- Orchestrate multi-component applications on standalone servers
- Deploy applications cluster-wide in container orchestrators
- Solve common deployment problems such as persistency or app exposure using best practices
- Review your application's health and debug it using open-source tools
- Discover how to orchestrate CI/CD workflows using containers

The Self-Taught Cloud Computing Engineer

Dr. Logan Song

ISBN: 978-1-80512-370-5

- Develop the core skills needed to work with cloud computing platforms such as AWS, Azure, and GCP

- Gain proficiency in compute, storage, and networking services across multi-cloud and hybrid-cloud environments

- Integrate cloud databases, big data, and machine learning services in multi-cloud environments

- Design and develop data pipelines, encompassing data ingestion, storage, processing, and visualization in the clouds

- Implement machine learning pipelines in a multi-cloud environment

- Secure cloud infrastructure ecosystems with advanced cloud security services

Packt is searching for authors like you

If you're interested in becoming an author for Packt, please visit `authors.packtpub.com` and apply today. We have worked with thousands of developers and tech professionals, just like you, to help them share their insight with the global tech community. You can make a general application, apply for a specific hot topic that we are recruiting an author for, or submit your own idea.

Share your thoughts

Now you've finished *Developing Blockchain Solutions in the Cloud*, we'd love to hear your thoughts! Scan the QR code below to go straight to the Amazon review page for this book and share your feedback or leave a review on the site that you purchased it from.

`https://packt.link/r/1837630178`

Your review is important to us and the tech community and will help us make sure we're delivering excellent quality content.

Download a free PDF copy of this book

Thanks for purchasing this book!

Do you like to read on the go but are unable to carry your print books everywhere?

Is your eBook purchase not compatible with the device of your choice?

Don't worry, now with every Packt book you get a DRM-free PDF version of that book at no cost.

Read anywhere, any place, on any device. Search, copy, and paste code from your favorite technical books directly into your application.

The perks don't stop there, you can get exclusive access to discounts, newsletters, and great free content in your inbox daily

Follow these simple steps to get the benefits:

1. Scan the QR code or visit the link below

https://packt.link/free-ebook/9781837630172

2. Submit your proof of purchase

3. That's it! We'll send your free PDF and other benefits to your email directly

Made in the USA
Monee, IL
07 July 2026